Lecture Notes in Economics and Mathematical Systems

626

Stefan Palan

Bubbles and Crashes in Experimental Asset Markets

Stefan Palan
Karl-Franzens-University Graz
Institute for Banking & Finance
Universitätsstraße 15/F2
8010 Graz
Austria
stefan.palan@uni-graz.at

The publication of this book was financially supported by the Karl-Franzens-University Graz and the government of the province of Styria, Austria.

ISSN 0075-8442
ISBN 978-3-642-02146-6 e-ISBN 978-3-642-02147-3
DOI 10.1007/978-3-642-02147-3
Springer Heidelberg Dordrecht London New York

Library of Congress Control Number: 2009928305

Cover design: SPi Publishing Services

Printed on acid-free paper

Springer is part of Springer Science+Business Media (www.springer.com)

Acknowledgement

I owe the greatest debt of gratitude to my fiancée Nicole Höhenberger, for her endurance, her uplifting words, her time spent proof-reading papers and this text, and for being herself. For his advice and support in all steps necessary on the path to become the researcher I want to be, I thank my supervisor Peter Steiner. My colleague Roland Mestel receives my warmest thanks, for daily advice, unmeasurable conceptual help and an open ear at all times. Ulrike Leopold-Wildburger deserves my heartfelt gratitude not only for offering to dedicate her time to being the second supervisor of this work, but also for setting me a deadline for the first experiment, thus probably accelerating this project by months. Alexander Brauneis had a big part in the success of this project, both by organizing two experimental sessions at the University of Klagenfurt and by showing me around town (and the best restaurants) while I was there. Antje Lutz' proof-reading has had a great impact in improving this text's quality and readability.

A special thanks goes to Catherine Eckel and Simon Gächter, for inspiring and encouraging me to take up experimental research during conversations at their seminar at the European Forum Alpbach 2005 and Simon's presentations at the Graz Schumpeter Lectures 2006. For the original idea of employing digital options in a Smith, Suchanek and Williams (1988)-type experiment, I thank Hubertus Hofkirchner. My thanks furthermore go to Lucy Ackert, James Ang, Ron King, and Erik Theissen, for help and invaluable comments, to Dave Porter for providing detailed data on Porter and Smith (1995), to Dan Gode and Shyam Sunder for providing detailed data on Gode and Sunder (1993), to Glenn Harrison for help with a tough-nut Stata problem, to Ernan Haruvy for information on Haruvy et al. (2007), and to Stefan Schmid and Urs Fischbacher for their help with the z-Tree software.

Last but not least, I wish to thank the participants of the CFS Summer School on Empirical Asset Pricing 2006 and the participants of the Mannheim Empirical Research Summer School 2008 for both advice and help regarding my research, and for the great time we had together. Finally, my thanks also go to Peter Lückoff, for making me aware of a valuable paper, to Volkmar Lautscham, Jonas Krotzek and Ulrich Pferschy for their mathematical help, and to Daniela Senkl and Franz Christian Wieser for their research assistance.

The financial support of the Department of Social and Economic Sciences at the Karl-Franzens-University Graz and the moral support of my colleagues at the Institute of Banking & Finance is gratefully acknowledged.

Contents

List of Abbreviations

The following abbreviations are used throughout the text:

AMEX	Acronym for the American stock exchange, a stock exchange based in New York City.
CAPM	Capital asset pricing model.
CBOE	Chicago board options exchange.
CRSP	Center for research in security prices at the university of chicago's graduate school of business.
DA	Double auction.
DAX	"Deutscher Aktienindex", the main index for the German equity market. The DAX contains the 30 largest companies of the Frankfurt stock exchange.
IMF	International monetary fund.
NASDAQ	Acronym for the national association of securities dealers automated quotations, a stock exchange based in New York City.
NBER	National Bureau of Economic Research in Cambridge, MA.
NYSE	Acronym for the New York stock exchange, a stock exchange based in New York City.
OTC	Over the counter. Refers to the trading of financial contracts directly between two parties, outside conventional exchanges.
PQ	Sealed bid-offer auction.
PQv	Tâtonnement version of the sealed bid-offer auction.
P(Q)	Variable quantity sealed bid-offer auction.
P(Q)v	Tâtonnement version of the variable quantity sealed bid-offer auction.
S&P 500	Standard & poor's index of 500 common stocks.
SUR	Seemingly unrelated regression.
TSU	Total stock of units, the total number of shares outstanding in an asset market experiment.
US, USA	The United States of America.

List of Symbols

The following symbols are used throughout the text:

$\mathrm{E}_s[\cdot]$	Subjective expectations operator of subject s.
$d_t, d_{r,t}$	Dividend per share in period t of round r.
f_t	Fundamental or dividend holding value in period t.
$f_t^{\min}, f_t^{\max}$	Minimum (maximum) possible remaining dividend payoff from one share from period t until the end of the round.
i_t	Transaction i in period t.
I_t	Total number of transactions in period t.
$\mathbb{M}^{\mathrm{DO8}}, \mathbb{M}^{\mathrm{DO5/10/15}}$	Set of option maturity dates in treatment DO8 and DO5/10/15, respectively.
$\mathbb{O}_{r,t=M}$	Set of options with maturity M in round r held by a subject, where $M \in \mathbb{M}^{\mathrm{DO8}} = \{8\}$ is the option's maturity date in the DO8 treatment and $M \in \mathbb{M}^{\mathrm{DO5/10/15}} = \{5, 10, 15\}$ in the DO5/10/15 treatment.
$OI_{r,t}$	Investment in options entered by a subject in period t of round r.
$p_{t=M}$	Stock price at the option maturity date.
P_{i_t}	Transaction price of transaction i in period t.
$\bar{P}_t$	Mean transaction price in the stock market in period t.
PO	A subject's total payout.
$PO_{t=M,\theta}$	Payoff from an option at maturity.
$PO_{t=M}^{o\in\mathbb{O}_{r,t=M}}$	Sum of the payoffs from all options in the set $\mathbb{O}_{r,t}$ when t equals the option's maturity M, for all M (and zero when t is not an option maturity date).
$Pr_s(\cdot)$	The subjective probability operator for subject s.
q	Total number of shares outstanding in the round (TSU).
q^r	Total number of transactions in round r.
q_t	Total number of stock transactions in period t.
r	Round in an experimental session.
s	Subject in the experiment.
R	Total number of rounds in a session.

SI	Stake invested into an option by each counterparty.
S_M	Stock price at an option maturity date.
$ST_{r,t}$	Sum of all proceeds from stock sales minus cost of all stock purchases in period t of round r.
t	Period within a round r.
T	Number of periods within a round r.
$x_{i_t}^{max}, x_{i_t}^{min}$	Binary variables which indicate whether transaction i in period t was conducted at a price exceeding (below) the maximum (minimum) possible dividend value of one unit of the experimental asset.
$x_{r,t}$	Number of shares held by the subject at the end of period t of round r.
x_t^{over}, x_t^{under}	Binary variables which indicate whether the mean transaction price in period t exceeded (was lower than) the maximum (minimum) possible dividend value of one unit of the experimental asset.
X	Strike price of an option.
$W_{r,t=0}$	Subject's initial wealth at the beginning of the first period of round r (including a possible loan).
Π_s	Subject s's profit from an option.
$\hat{\sigma}_{P_{i_t}}$	Sample standard deviation of transaction prices in period t.
σ_{f_t}	Ex ante standard deviation of the ex post fundamental value of the asset in period t.
θ	Binary variable equaling unity if the option a subject holds is a digital call and zero if it is a digital put.

List of Figures[1]

Appendix

[1]All figures in this book were created by the author, unless otherwise noted.

List of Tables[2]

List of Examples[3]

[2]All tables in this book were created by the author, no declarations regarding the source of individual tables will therefore be given.
[3]All examples in this book were created by the author, no declarations regarding the source of individual examples will therefore be given.

Chapter 1
Introduction and Motivation

The beginning of knowledge is the discovery of something we do not understand.

Frank Herbert, 1920–1986

1.1 Prediction Markets and Online Betting Sites

Since the beginning of human history, trade has been an essential part of the lives of the human race and has played an important role in the development of civilization. Very early, centralized places of exchange were established to facilitate trade, promote competition, and reap efficiency gains.[1] All of these marketplaces had in common that transactions could only be undertaken when the counterparties were physically present at the location of the marketplace. This changed only in the last century, with the invention of first the telephone and later the computer. Today, most stock exchanges are either fully electronic or in the process of transitioning to such a state. The proliferation of the internet and web-based applications laid the groundwork for the facilitation and geographic dispersal of market transactions. This has led to an enormous expansion of the set of possible traders, yet had limited impact on trading overall. The reason behind this was the remaining requirement at most exchanges for trades to go through brokers, which – combined with the still substantial transaction costs – limited the set of traders mostly to investment professionals. The advent of online betting sites catering to private users alleviated these constraints, opening the door to a price discovery process that rapidly includes information of private investors.

[1]Tremel (1969) for example wrote about trade as early as in the neolithic, and mentions fortified marketplaces in the bronze age. See also Walter (2006) and Lowry (2007).

S. Palan, *Bubbles and Crashes in Experimental Asset Markets*, Lecture Notes in Economics and Mathematical Systems 626,
DOI: 10.1007/978-3-642-02147-3_1, © Springer-Verlag Berlin Heidelberg 2009

These developments also form the first component motivating this research effort – the emergence of a new type of online marketplace, making information regarding financial market prices a good that is the underlying for financial contracts. Prediction markets and online betting sites like binarybet.com, ideafutures.com, intrade.com, mybet.com, newsfutures.com, and redmonitor.com give investors an outlet to trade on information very cheaply, with practically no barriers to entry, and carry the additional advantage of operating around the clock (with occasional trading stops for system maintenance). Some of these sites employ cash-or-nothing (digital) options as their vehicle of trade, which is the reason why this financial instrument was accorded central attention in this book. A cash-or-nothing option pays out a fixed cash amount in the case that it expires in the money and nothing if it expires out of the money.[2] This payoff pattern is in some ways superior to that of standard options, in that it is conceptually easier to understand, since it resembles the payoff from an everyday bet.[3] This fact might improve the adoption of such options by non-sophisticated investors, a conjecture that is supported by the large number of traders in these markets betting not only on financial market prices, but also on such varied topics as the success of movies, election outcomes, and the date of anticipated scientific breakthroughs.

1.2 Bubbles and Crashes in Financial Markets

The second important piece of motivation for this research project is the propensity of market prices to sometimes exhibit extraordinary run-ups (bubbles) followed by crashes back to levels closer to fundamental values. Such bubbles and crashes in financial markets are no phenomenon unique to modern financial systems or highly interconnected marketplaces. Rather, they have been documented as early as after the disintegration of the tulip price bubble in the Netherlands in 1637 or the plunge in stock prices of the South Sea Company in the UK in 1720.

---

[2]A digital option contract – as the term is used in this text – can be thought of as a separate digital put and a digital call option. If the price of the underlying at maturity exceeds the strike price of the option, the call (put) option part is *in the money* (*out of the money*), with the reverse being true if the price of the underlying is lower than the strike price at maturity. The party whose position is in the money receives a fixed payoff which is the sum of the (equal) stakes invested into the option by the two contract partners at its inception. The second party receives nothing. In the case that the digital option is *at the money* at maturity, each contract partner sees her initial stake returned.

[3]Cp. Oliver (2007), pp. 127–128. Oliver (2007) reported on binary betting contracts in prediction markets, writing that these contracts were marketed as being very simple, because market participants only had to decide whether they want to bet on the market going up or down and because the maximum gain and loss were known in advance.

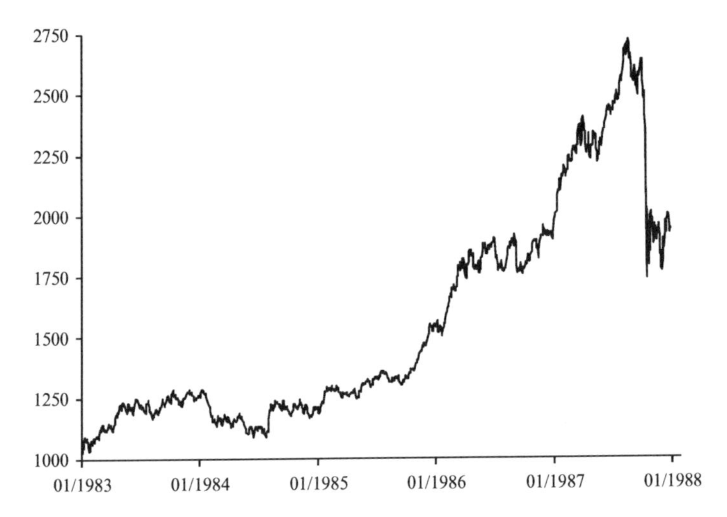

Fig. 1 Dow Jones Industrial Average 01/1983–01/1988. Dow Jones Industrial Average, daily price index, 01/1983–01/1988.
Source: Thomson Datastream: Dow Jones

Figure 1 above gives an example for such a price pattern that has ex post been labeled a bubble-and-crash pattern in a number of studies (e.g., West (1988), Ang et al. (1992), Caginalp et al. (2000a), Westerhoff (2003)).[4]

The impact of bubbles – and the attention they received in the economic discipline – has greatly increased with the growing interconnectedness of today's financial markets. What makes these phenomena so problematic is that bubble-and-crash patterns in financial market prices are widely considered harmful to economic activity. This can be traced to the importance of market prices for the allocation of investment capital.[5] Market prices form the basis for the allocation of investment capital to its most efficient uses in the real economy. This implies that a misallocation of available resources to non-optimal uses results for the case where market prices are less than perfect measures of underlying values.[6] In the previous century, the Great Depression in the 1930s clearly demonstrated the danger that spillover effects from price bubbles in financial markets pose for the underlying real economy. As a more recent example, Gan's (2007) study showcased an indirect transmission channel from asset market bubbles into negative effects for the real economy, underlining the possible efficiency gains to be had from a better understanding of the bubble phenomenon. In his study, he found that, in Japan in the early 1990s, the dramatic

---

[4]The careful wording is due to the fact that there are studies which dispute the bubble explanation given for a number of anomalous price patterns in the past. See e.g., Donaldson and Kamstra (1996), and Pástor and Veronesi (2006) for an alternative explanation of price bubbles in the 1920s and in the late 1990s, respectively. These two articles are briefly summarized in Sect. 2.1.3.

[5]Cp. e.g., Shiller (2003), p. 102.

[6]Cp. e.g., Friedman (1984a) for a brief discussion of the role market prices play in the allocation of scarce resources to economic ventures.

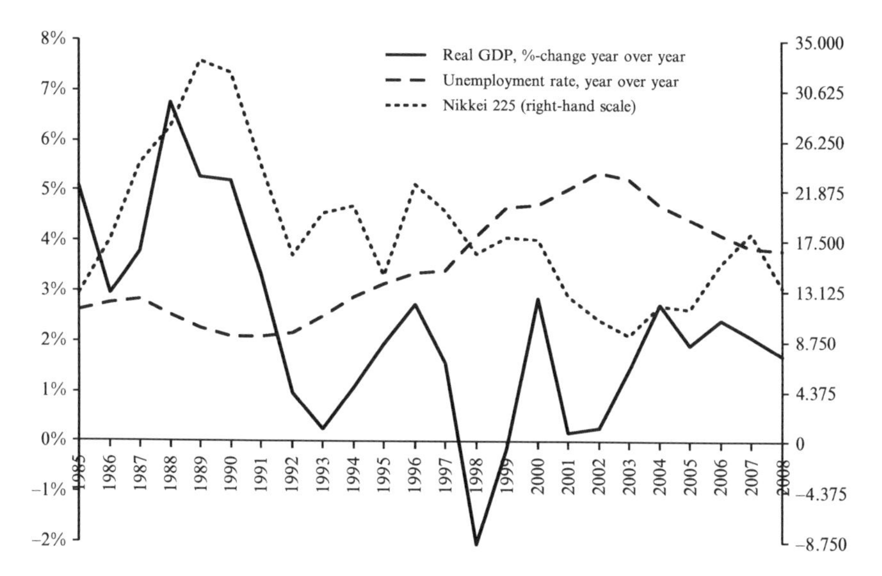

Fig. 2 Indicators of Real Economic Performance in Japan. The figure graphs the percentage change in Japan's real GDP, its unemployment rate, and the level of the Nikkei 225 equity index (right-hand scale), from 1985 to 2008.
Source: Thomson Datastream: OECD Economic Outlook (GDP and unemployment) and Nikkei (Nikkei 225)

drop in land prices of almost 50% led credit-constrained banks to reduce their lending activity by about one third, which in turn was responsible for about one fifth of the observed decline in fixed investment in the real economy and a quarter of the reduction in stock market valuation. Figure 2 below illustrates this impact through the main indicators of real economic activity – GDP and the unemployment rate – as well as through the level of Japan's main stock index, the Nikkei 225. It clearly shows the nearly parallel decline in GDP growth and stock valuation in the early 1990s, as well as the delayed response in the unemployment rate.

Examples like this one illustrate the possible benefits of reducing such anomalies in financial markets – a goal that this book hopes to advance, by helping the economic discipline in gaining a better understanding of the bubble phenomenon.[7]

The bursting of the bubble in real estate prices in the economy of the USA in the second half of 2007 is a more recent example of the repercussions extreme fluctuations in financial market prices can have on the real economy. At the present time, the United States are in a recession that was caused by the precipitous drop of real estate prices, which in turn led to a depreciation of the collateral held by banks for their loans. As more and more banks lost the trust of their customers and fellow financial institutes, they ran into liquidity constraints, forcing them to limit their lending activity. This in turn led to an increase in defaults on debt by borrowers who could not refinance their outstanding loans, which – in a continuation of this

[7] The literature on price bubbles is summarized in Sect. 2.1.3.

destructive domino effect – resulted in a reduction of private spending. The US economy's strong dependence on the American consumer, finally, led to the economic downturn that can be observed in US markets today. In previous centuries, such a crisis in the USA would have been limited to US markets. Due to the increased interconnectedness of financial markets that was mentioned above, its effects today spill over into the Asian and European economies, as the International Monetary Fund (2008) forecasted already in April 2008. This worldwide scope of local crises underlines the importance of inquiries into the causes and conditions necessary for bubble-and-crash patterns in financial markets.

1.3 The Role of Derivative Markets in Informational Efficiency

The third strand of research impacting on the choice of this book's topic saw its beginnings in the 1970s. Cox (1976) was one of the earliest articles to model the link between futures trading and the information processing taking place in the formation of spot market prices. Since then, an extensive branch of literature has been devoted to the connection between the trading of forwards, futures and options and its impact on the informational efficiency of the market prices of the underlying asset. Compared to many other topics in financial economics, the results are surprisingly unequivocal. Both theoretical and empirical studies of financial markets have shown that derivative markets generally should and do process information earlier and faster than spot markets and that the creation of a derivative market to accompany a spot market usually leads to higher price efficiency in the latter. One explanation for this effect was proposed by Figlewski and Webb (1993), who reasoned that options give traders who cannot or will not engage in short sales due to e.g., transaction costs, an opportunity to sell short indirectly. As is the case in a number of similar studies, they found that derivative markets are the primary trading venue for informed traders and therefore play a primary role in the incorporation of information into market prices.[8]

These results on the positive impact of derivative markets on the efficiency of the related spot markets forge the missing link in the chain of thoughts developed above: If derivative markets improve the informational efficiency of spot markets, can then prediction markets – which are just another form of a marketplace for the trading in derivative contracts – reduce or prevent the formation of price bubbles at financial exchanges? This question becomes all the more relevant considering that on the one hand, trading in online prediction markets requires lower capital than conventional derivatives markets, thus possibly attracting more diverse traders who are likely to hold more diverse information. On the other hand, these markets are

[8] A more detailed overview of the research into the role derivative markets play in information dissemination and incorporation is given in Sect. 2.2.

open around the clock for trading from any computer in the world. These facts suggest that prediction markets could possibly improve market efficiency in conventional markets. The transmission channels could be both the attraction of more (and more diverse) traders, and the lower transaction costs. The latter could make it viable to trade derivatives on stocks previously considered too small for other than over the counter (OTC) derivative trading. Yet it is exactly these relatively small and illiquid, "neglected" stocks which should profit the most from additional price discovery. OTC trading, because of its more localized nature, does not lead to a similarly widespread dissemination of the information contained in transaction prices as trading in conventional exchanges.

1.4 Methodology

This book brings together the above three pieces to form a jigsaw that can be summarized as follows: Bubbles in financial markets are an expression of market inefficiency that causes damage to the real economy. Given that derivative markets have been documented to improve information dissemination and incorporation into prices – do new trading ventures, using previously rare forms of financial derivatives and giving a broad base of traders access, improve market efficiency? Can the incidence and extent of price bubbles in financial markets be reduced if markets are provided with the forward-looking price information from digital option markets? Since high-quality price information is very valuable for both corporate finance (M&A, hedging, valuation, etc.) and financial markets (derivative pricing, possible reduction of the frequency and size of bubbles), improved pricing information from prediction markets could – contingent on their success – lead to a gradual balancing reaction of securities markets. Through such a process, that information would become reflected in prices, thus improving these markets' informational efficiency. Unfortunately, such a balancing reaction by financial markets is bound to be slow and its empirical detection would be hampered by the noise that is necessarily present in financial market prices. For this reason, the research questions were explored using techniques from the field of experimental economics. Laboratory experiments, apart from being replicable and allowing for the variation of experimental treatments,[9] have the advantage of permitting the experimenter to specify a known fundamental value process – information that cannot be observed in most real-world markets.

The experiment reported in this book is loosely based on the seminal work of Smith et al. (1988). Their article sparked a number of studies investigating the causes and properties of bubbles in experimental asset markets. In their baseline market,

[9] A treatment is one set of design parameters in an experimental study. For a clear definition of this and other terms employed in the description of experimental work, refer to the first paragraphs of Chap. 3 Note that in the literature, the same terms may sometimes refer to different underlying concepts.

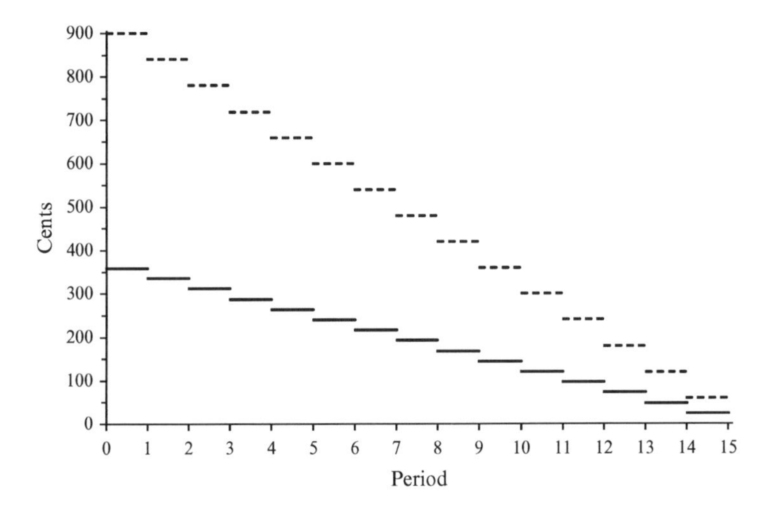

Fig. 3 Dividend Value of One Unit of the Experimental Asset. The solid (broken) line plots the expected (maximum) sum of future dividend payments during the life of an asset with an expected (maximum) dividend payout of 24 (60) at the end of each period, over 15 periods

groups of nine to twelve students participated in one to three repetitions of the same 15-period market. Each subject was endowed with experimental currency and shares of an unspecified asset, which could be exchanged for each other in an electronic double auction market. Each period was followed by a payout of a random dividend for each unit of the asset traders held in their possession. This dividend was discretely and uniformly distributed and could have one of four possible non-negative values in each period. The fundamental value of a unit of the asset in their experiments was common knowledge and declined deterministically to zero over time, as plotted in the solid, stepwise decreasing function in Fig. 3.[10]

Despite this last fact, Smith et al. (1988) observed large deviations of transaction prices from the fundamental value, forming bubbles which in some cases even exceeded the maximum possible value the asset could return in dividends (in the case where only the highest dividend would be drawn in each future period), shown as the broken stepwise decreasing function in Fig. 3 (the asset's terminal value was zero). However, once subjects gained experience by participating in repeat rounds, they tended to converge on rational, common, intrinsic dividend value expectations.

Smith et al.'s (1988) results were replicated numerous times by a number of studies and turned out to be impressively robust to various treatments (cp. Sect. 2.4 for a detailed presentation and comparison of these studies). One article in which the authors reported having achieved an improvement in the propensity of spot market

[10]Figure 3 plots the fundamental value and maximum value of one unit of asset for an asset with an expected dividend payout per period of 24 and a maximum dividend per period of 60. Smith et al. (1988) used various dividend payoff regimes, but this is the one that was used most often in subsequent studies. Nonetheless, the general pattern fits all of their treatments, as they always used a discrete uniform distribution with four possible and equiprobable non-negative payouts.

prices to exhibit an inefficient price bubble was Porter and Smith (1995). Porter and Smith tested a market design where, in addition to the spot market for an unspecified, dividend-paying good, they enabled the trading of futures contracts on that good. They conjectured that the possibility to trade on the asset's price in the future would facilitate a mechanism of backward induction, leading subjects to refrain from trading the experimental asset at inflated prices. Porter and Smith (1995) reported that the futures market reduced the bubble amplitude and had no significant effect on duration and turnover of the bubble with inexperienced traders in the futures market, but exhibited significantly reduced turnover with experienced futures traders. They interpreted their findings to signify "that an important function of a futures market is to reduce each individual's uncertainty about other peoples' [sic!] expectations."[11]

The findings from this experiment are consistent with the literature investigating the impact of derivative markets on the informational efficiency of spot markets that was briefly introduced above and is described in more detail in Sect. 2.2. It formed the starting point of the methodological approach followed in the present study. This approach consisted of programming an experimental asset market environment similar to that employed in Smith et al. (1988), and complementing it with a digital option market, following a similar design as the futures market in Porter and Smith (1995). The aim of this setup was to determine through experiments conducted with business and economics students[12] whether the possibility of trading digital options – in addition to trading in the spot market – reduces the occurrence of bubbles in financial spot market prices.

1.5 Scientific Relevance

The research question driving this research effort can be located at the confluence of the informational efficiency of financial markets with dimensions like market structure, expectation formation and individual behavior. More specifically, the study reported herein aimed to analyze one possible way to improve market efficiency – the opening of derivative markets which serve to increase information flow and complete spot markets. Over time, however, this focus changed, as the author recognized that the description of the expectation formation process followed by subjects in this type of experimental market was more central to the appearance of bubbles than the exact form of the market institution. It was found that the root of the bubble phenomenon in Smith et al. (1988)-type markets lies less

[11] Porter and Smith (1995), p. 525.

[12] Using student subjects is standard practice in experimental economics. Studies investigating the impact of differing subject pools in economics experiments frequently find effects on experimental outcomes when employing students who have little to no prior financial knowledge, yet usually report finding no difference between the results from experiments with business and economics majors and real business professionals. Section 2.4.4.6 contains a summary of studies investigating the impact of using subjects from different socio-demographic and educational groups.

in specific forms of exchange or other characteristics of the exchange framework than in the learning strategy and the cognitive processes employed by individuals.

Up until the last decades of the previous century, economics relied on (neo) classical approaches to explain phenomena encountered in real-world markets. With the advent of behavioral finance – and, to a good part, experimental economics – economists increasingly started to consider other than classical explanations, especially for observations that run counter to classical theory. The bubble-and-crash pattern in experimental asset markets of the type introduced by Smith et al. (1988) is such a case, where little progress was made in more than 20 years of research based on the assumption of rational agents. Recently, Lei et al. (2001) established a breakthrough result when they proved that this pattern can be fully explained *only* by a theory containing irrational agents. Shifting the research focus of the present text away from the specific market institution employed and toward the behavioral explanation for the observed price patterns as well as toward the role rational and irrational traders play in this setting bore fruit in the form of a new theory of how subjects form expectations in Smith et al. (1988)-type markets. This theory will, later in this book, be referred to as the Hypothesis of Observational Belief-Adaptation, a hypothesis of how subjects in such a situation, even though they are not rational, learn to act in such a way that their actions produce a price pattern closely mirroring that of rational agents. It succeeds in describing a number of observations that cannot easily be reconciled with the assumption of rational agents. More importantly, it further allows for the formulation of new research questions which can provide additional illumination regarding its validity and the limits of its applicability in differing settings.

In December 2007, Ernan Haruvy, Yaron Lahav and Charles Noussair published an article in the *American Economic Review*, which underlined the scientific relevance of this result. Haruvy et al. (2007) had conducted market experiments of the type invented by Smith et al. (1988), wishing to address the question of "how expectations evolve, respond to, and influence a market as it converges to fundamental pricing."[13] Their findings on adaptive beliefs in these markets accord nearly perfectly with the Hypothesis of Observational Belief-Adaptation formulated in Chap. 4.2.1. This corroborates – via no less than a publication in the *American Economic Review* – the scientific relevance of the research topic.

Seen from a macro perspective, the relevance of these findings derives from the better understanding it gives the economic discipline of individuals' actions in financial markets, a topic that has been studied for hundreds of years (cp. e.g., Smith (1843)). It represents an important step, in that it explains why and how bubbles form in a market of the Smith et al. (1988)-type. While the metaphorical goal might be nearer now, however, it has not yet been reached. Necessary steps for the future include using this new knowledge to find a way to prevent bubbles in this type of market, and generalizing this solution to real-world financial markets. Only when – and if – these steps are achieved, can the threat the bubble-and-crash phenomenon constitutes for financial markets be considered banished.

[13] Haruvy et al. (2007), p. 1902.

Chapter 2
Literature Review

It takes a great piece of history to produce a little literature.
Henry James, 1843–1916

2.1 Literature on Market Efficiency[1]

The role of bubbles in financial markets is intricately connected to the question of informational efficiency. The reason is both that bubbles above and below fundamental values are a violation of market efficiency, and that the fundamental value itself and deviations from it can only be defined with reference to a framework of informational efficiency in a market (cp. Roll's critique in Roll (1977)). Because of this observation, this section starts with a short introduction to the topic of market efficiency (Sect. 2.1.1 below), briefly reviews evidence of market inefficiency (Sect. 2.1.2), and finally spends some time on the specific anomaly of price bubbles (Sect. 2.1.3).

2.1.1 Literature in Favor of the Efficient Market Hypothesis

If there is to be one "father" of the efficient market hypothesis, this man is Eugene Fama, who remains an outspoken proponent of the hypothesis to this day. In Fama (1970, 1991, 1998), he gave comprehensive overviews of the literature on the topic

[1] As one of the best-researched topics in modern finance, the efficient market hypothesis has been the subject of countless papers and it would exceed the scope of this text to give a more comprehensive overview than the brief introduction in this section. The interested reader is referred to Palan (2004) for a more extensive discussion of the literature on market efficiency.

S. Palan, *Bubbles and Crashes in Experimental Asset Markets*,
Lecture Notes in Economics and Mathematical Systems 626,
DOI: 10.1007/978-3-642-02147-3_2, © Springer-Verlag Berlin Heidelberg 2009

and documented its evolution over the three decades spanned by these papers. Fama defined an efficient market as "A market in which prices always 'fully reflect' available information,"[2] and proposed the classifications of weak-form, semistrong-form, and strong-form market efficiency to concretize the "available information." These three categories have by now become the standard in descriptions of market efficiency.

Nonetheless, the history of the efficient market hypothesis had begun earlier. Bachelier (1900)[3] laid the theoretical groundwork for the efficient market hypothesis, which was postulated half a century later by Maurice Kendall. Kendall (1953) found that stock prices evolved randomly and that his data offered no way to predict future price movements. The explanation for this phenomenon, the efficient market hypothesis, initially seemed counterintuitive to the academic community. However, after the first shock had passed, scholars quickly embraced the theory and began to document its validity in real-world markets by studying empirical data.

To do so, they developed different frameworks to model the characteristics of market prices. The first type of framework – based on expected return efficient markets – includes such well-known models as the fair game model, the random walk and the submartingale models, as well as the market model and the famous capital asset pricing model (CAPM) of Sharpe (1964); Lintner (1965); Mossin (1996).

In the years from the 1950s to the 1970s, most studies based on the CAPM and fair game models found evidence consistent with the efficient market hypothesis. Despite some evidence to the contrary from the variance-based literature (which will be introduced below), by the early 1970s markets had therefore largely come to be considered to be efficient in the semistrong form, as defined by Fama (1970). As a case in point, Malkiel noted with regard to market efficiency:[4]

> "I don't know of any idea in economics that I've studied and been associated with over this period of time [since the first publication of 'A Random Walk Down Wall Street' in 1973] that has held up as well."

A second class of models used to test market efficiency focuses on variance as the key characteristic. Among them are the model of Shiller (1981), who reported that stock prices were too volatile to be efficient when compared to subsequent dividend payouts, and the model of Marsh and Merton (1986), which showed that Shiller's results could be reversed by a change in assumptions regarding the dividend model. The reply of Schwartz (1970) to the seminal paper of Fama (1970) could also be considered to fall into the category of variance efficient market models, as it propagated the use of models that tested for variance-based strategies to generate excess returns in capital markets.

The first variance efficient market models in the early 1980s coincided with the advent of behavioral finance and behavioral market models, which soon started to

---

[2] Fama (1970), p. 383.

[3] As quoted in Ziemba (1994), p. 200.

[4] Malkiel et al. (2005), p. 124.

erode the solid standing the efficient market hypothesis had (until that time) enjoyed in academic circles.[5] A number of anomalies were discovered in empirical data, suggesting that the universal belief in the applicability of the efficient market theory had been overly optimistic. Today, evidence of widespread efficiency in developed markets coexists with well-recognized anomalies, both in these highly developed markets in industrialized countries and – much more frequently – in less developed market economies. These anomalies can be subsumed under a few broad categories, which are summarized in the following section.

2.1.2 Literature on Market Inefficiencies and Anomalies

Over the years, a substantial number of market inefficiencies or "anomalies" has been documented. Among them are the serial correlation of returns and variances, return seasonality, the neglected-firm and liquidity effect, and excess returns earned by insiders. The following paragraphs give a brief overview of this literature, which is reviewed in more detail in Palan (2004). Due to its prominent relevance for the present study, the literature on asset price bubbles is discussed separately in the next section.

In certain instances, securities have been found to display autocorrelation of returns and of return variability – a topic that has received considerable attention since the 1990s. Such a property of time series of returns indicates a lack of market efficiency, since the inequality of conditional and unconditional expectations violates the fair game model of financial market returns. The search for serial correlation in these variables is probably the most straightforward test for market efficiency, although shortcomings of the measurement techniques often cast doubts on the validity of results. The anomaly of serial correlation is in the literature frequently referred to as a "short-term momentum, long-term reversal" effect.[6] The reason for this moniker is that early studies detected evidence of positive serial correlation (i.e., momentum) over periods of up to 12 months, while finding negative serial correlation (i.e., reversal) for periods ranging from 13 to 60 months. Conrad and Kaul (1988) for example reported positive serial autocorrelations for stocks listed at the New York Stock Exchange (NYSE). Jegadeesh and Titman (1993) found both short-term momentum and long-term reversal for stocks from the database maintained at the Center for Research in Securities Prices at the University of Chicago (CRSP), and De Bondt and Thaler (1985) documented negative serial correlation for the same underlyings. Negative serial correlation over longer time periods is also a result of the studies by Fama and French (1988), and Poterba and Summers (1988). Rouwenhorst (1998) extended the analysis to twelve European countries, finding a similar momentum-and-reversal effect for his 1978–1995

---

[5]Some selected papers of this strand of the literature are Black (1986); Shleifer and Summers (1990) and – for a more critical view – Fama (1998).

[6]Cp. e.g., Jegadeesh and Titman (1993).

sample. Later studies, however, provided evidence that this effect might be decreasing or disappearing over time (e.g., Jegadeesh and Titman (2001)), or disputed its presence altogether (Fama (1998)).

Anomalies subsumed under the heading of return seasonality are characterized by patterns in financial asset prices or in their variability that recur regularly at specific calendar dates and times. Seasonality has been documented in intraday, weekly, monthly and annual return data. A famous example of such a pattern is the day-of-the-week effect or weekend effect – the observation that returns at the beginning of a week are more likely than not to be below average, while returns at the end of the week are frequently higher than the average. Studies documenting this phenomenon are e.g., Cross (1973); French (1980); Gibbons and Hess (1981); Keim and Stambaugh (1984). Similarly well-known is the turn-of-the-year effect, January effect, or the small-firms-in-January effect, which refers to the pattern that returns tend to be higher in January than over the rest of the year, particularly for small firms. (Cp. e.g., Keim (1983); Rogalski (1984); Ziemba (1988); Ritter and Chopra (1989).)

The neglected-firm effect and the effect of stock prices' reaction to the inclusion of a stock into an equity index can be subsumed under the heading of liquidity effects. The former was coined by studies which showed that, compared to larger firms, small and less-reported-on firms offer a liquidity premium, because investors purchasing them are subject to liquidity risk (cp e.g., Amihud and Mendelson (1986, 1991); Pratt (1989); Chordia et al. (2000); Ross et al. (2005)).The second term refers to a finding by Shleifer (1986), who studied the price reaction stocks exhibited upon being included into a market index. A stock's index inclusion is an event that arguably does not reveal new information about the stock, but does cause purchases by mutual funds, which are in many cases accompanied by a liquidity crunch with a concurrent effect on prices.

Finally, the evidence on the question of whether individuals privy to inside information can earn excess returns (i.e., markets not immediately adjusting to inside information) is relatively unequivocal. It was confirmed in studies like Pratt and DeVere (1968); Jaffe (1974); Lorie and Niederhoffer (1968); Seyhun (1986). In a rare conflicting result, Hawawini (1984) found evidence consistent with strong-form market efficiency for French, Spanish and U.K. mutual funds.[7]

2.1.3 Price Bubbles

Bubbles in financial market prices have already been briefly discussed in Sect. 1.2. They are a sign of inefficient markets, because they lead to an inefficient allocation of capital to productive uses. Bubbles are a phenomenon that has received relatively

[7] However, Hawawini relied on the assumption that mutual fund managers possess insider information. If they do not, his evidence lends support only to semistrong-form efficiency.

widespread attention compared to other signs of market inefficiencies, which might be due to an issue of magnitude: Most findings of inefficiencies in market prices are small; so small in fact that they are often only statistically – but not economically – significant. The same does not apply to bubbles, which in the form of stock (or, more recently, real estate) market crashes received attention not only in the financial but also in the mainstream press.[8] Naturally, science also took up the topic both in theoretical and empirical work, some of which is summarized below. Note that bubbles seem to be a research subject with a particularly bright future, since scientists cannot only not agree on what exactly causes bubbles, but rather hold differing opinions even on the question of whether stock market prices in the late 1990s and early 2000's, or in the great depression, could actually be considered bubbles. The reason for this lack of agreement lies both in problems of measurement and statistical technique and in the different definitions used by different scholars. To shed some light on this literature, the following paragraphs list a number of bubble definitions, discuss their differences and present the literature dealing with this phenomenon.

As one of the early papers dealing with bubbles in a theoretical model, Diba and Grossman (1988) defined a *rational bubble* as follows:[9]

> "A rational bubble reflects a self-confirming belief that an asset's price depends on a variable (or a combination of variables) that is intrinsically irrelevant–that is, not part of market fundamentals–or on truly relevant variables in a way that involves parameters that are not part of market fundamentals."

This argument is reminiscent of the *sunspot* literature, which is captured well in the seminal article by Cass and Shell (1983). A sunspot is – in the words from above – a variable that is intrinsically irrelevant, yet influences prices nonetheless.[10]

Camerer (1989) found that what he calls *rational bubbles* can occur if rational traders expect to profit from participating in the bubble. He points out that under *common knowledge of rational expectations*, each trader should expect to on average make a loss by purchasing at excessive prices, because the *average* trader cannot expect to resell the asset at an even higher price, and each trader is equally likely to be in the losing group. This is because common knowledge of rational expectations implies an infinite conditioning on others' information, in that each trader knows that each trader knows that each trader knows ... that all traders in the market are rational, which is a sufficient condition to ensure that prices follow fundamental values and do not exhibit even rational bubbles.

Assuming rational traders but no common knowledge of this fact, the ingredient missing for a bubble in Camerer (1989) is a departure from rationality, for which he

---

[8] See e.g., Independent (2001), International Herald Tribune (2007); New York Times (2008). A prescient article regarding today's housing crisis was for example Los Angeles Times (2005).

[9] Diba and Grossman (1988), p. 520.

[10] Kraus and Smith (1998) define a *pseudo-bubble* as a bubble based on sunspots, with prices which stay above or below fundamental value over all trading dates. Since this type of bubble is of no particular relevance for this book, however, it will not be discussed here in more detail.

suggested overconfidence as a natural candidate. Overly optimistic expectations are a well-documented trait of the human species,[11] which Camerer argued is rational if it has biological (i.e., evolutionary advantage for optimistic individuals) or psychological (i.e., preference for optimistic belief) value. Furthermore, what Camerer called *near-rational bubbles* are possible if traders are unsure about others' beliefs and perceive a positive (subjective) probability that other traders will expect a given bubble to burst at a later point in time than the time at which they themselves expect it to burst. This argument is reminiscent of the *winner's curse* phenomenon[12] in that the individuals with the largest positive error term in the estimation of the time until the bubble bursts are the most likely to end up holding the overvalued asset when the bubble does indeed burst. As in the case of the winner's curse, individuals in such a situation should adjust their expectations to take account of this fact but – just like there – often fail to do so. A complicating factor in this dilemma are the dynamics of the problem: In the winner's curse, an individual is "cursed" if she ends up purchasing an asset at a price above its (ex ante unknown) fundamental value. Yet, in that setting, the individual could evade this problem by adjusting her value expectations downward. In the bubble example, this is only possible ceteris paribus, but not if all other market participants likewise adjust their expectations. If they do so in a rational way, their backward iterative reasoning will step-by-step lead them to (cognitively) reduce the length of the bubble period, until it finally disappears entirely, causing the inflated market prices to drop immediately. Even if investors are only partially rational, it is hard to see by which amount one should revise one's expectation of the bubble's length, when that very number depends on the expectations and revisions of all other agents.[13]

Allen and Gorton (1993) proposed a theoretical model to similarly show that settings can exist where rational behavior is consistent with stock price bubbles. The novelty of their approach was to populate the model with – among others – portfolio managers, who pick stocks for investors, but have only limited liability. Their position is that of a call option, which makes them willing to buy stocks which are overvalued, if there is a positive probability that prices will increase further

[11] See e.g., Svenson (1981) for evidence of overconfidence among automobile drivers, Roll (1986) for displays of overconfidence among managers, and Camerer (1987) for overconfidence among experimental subjects.

[12] See e.g., Wilson (1977) and Milgrom and Weber (1982).

[13] This observation might remind the reader of another famous example from the economic literature – that of the p-beauty contest of Moulin (1986), which in turn derives from Keynes' (1936) famous beauty contest. In Moulin's example, the task was to pick, out of the interval from 0 to 100, a number that comes as close as possible to 2/3 of the average of all numbers submitted. Naturally, like in the bubble problem above, this leads to an infinite conditioning, where one tries to pick the number that is 2/3 of the number the average person thinks is 2/3 of the number the average person thinks is 2/3 the number the average person thinks ... the average person will pick. In both the bubble and in Moulin's example, zero is the rational solution for the length of the bubble period and the number to pick, respectively. However, in both examples, the evidence suggests that the average individual does not act rationally and expects (picks) a bubble of positive length (a positive number).

before they need to sell. While otherwise plausible, the model unfortunately relies on exogenously determined, monotonously increasing security prices – a feature that renders the model rather unrealistic and limits the conclusions which can be drawn from its outcomes.

A different bubble definition is used in the theoretical model of Allen et al. (1993), where an *expected bubble* occurs whenever the price strictly exceeds each agent's expected value of the asset. A *strong bubble*, in turn, is defined as a price where every agent knows that it strictly exceeds the possible future dividends.[14] Figure 3 on p. 7 illustrates these concepts. An expected bubble as defined by Allen et al. (1993) would be characterized by prices lying above the solid line, while in a strong bubble prices would exceed even the broken line. Allen et al. (1993) found that – in their rational expectations model – necessary conditions for the existence of expected bubbles are ex ante inefficient endowments, and a short-sales constraint for every agent in some state of nature at a time later than that at which the bubble occurs. Furthermore, for strong bubbles, all agents must also have some private information that is not revealed in equilibrium prices, and their actions must not be common knowledge.

De Long et al. (1990) probed the role of rational speculators in markets characterized by positive feedback traders. In their model, rational speculators buy stock following price increases. Once feedback traders catch on to the trend of increasing prices and start buying themselves, the rational speculators sell their holdings and reap capital gains. By mimicking the actions of positive feedback traders, rational speculators in their model destabilize prices and increase overvaluations.[15] This behavior of the two heterogeneous groups of traders leads to positive autocorrelation of returns in the short run and negative autocorrelation in the long run, a pattern that conforms well to the short-run momentum and long-run reversal effect reviewed in Sect. 2.1.2 above. Furthermore – as De Long et al. (1990) pointed out in their motivation – their findings are consistent with accounts of the investment strategies of investors like George Soros and others, as well as with the intent behind market newsletters and some investment pools.

Moving away from theoretical models and toward empirical work, Guenster et al. (2007) analyzed bubbles in the context of US industries, using the CAPM, the Fama and French (1993) model, and the Carhart (1997) model to derive fundamental values. Defining bubbles as price patterns where the price's growth rate exceeds that of fundamental value and where the growth rate of price experiences a sudden acceleration, they found a significantly positive relation between the occurrences of bubbles and subsequent abnormal returns of between 0.41% and 0.64%. On the other

[14] Actually, theirs is a three-period model with a single liquidating dividend, so they formulate their definition as follows: "We will say a 'strong bubble' exists if there is a state of the world such that, in that state, every agent knows (assigns probability 1 to the event) that the price of the asset is strictly above the liquidating dividend." (Allen et al. (1993), p. 211) For the sake of this book, their definition is generalized to the case where there is more than one future dividend – as stated in the text above. Note that this definition is silent on the role of discounting.

[15] Note that this mechanism describes closely observations made during the course of the experiments conducted for this book, which are discussed below in Sect. 4.2.3.

hand, bubbles were accompanied by a doubling of the probability of a crash (defined as a return below 1.65 times the standard deviation of abnormal returns) in subsequent months. Nonetheless, their results indicated that the additional risk upon detection of a bubble was more than outweighed by the prospect of superior returns in their sample. Finally, they reported that, conditional on a crash having occurred in the preceding 12 months, another crash became more likely during the following months.

A counterpoint to the majority view of bubbles being present in the world's stock markets is formed by articles like Donaldson and Kamstra (1996), and Pástor and Veronesi (2006). The former showed that dividend forecasts in the 1920s justified the stock prices prior to the market crash in 1929, while the latter demonstrated that the high expectations with regard to the riskiness of NASDAQ stocks in the 1990s suggest that the observed prices prior to the sharp decline in the early years of the twenty first century had been justified. On another note, Barlevy (2007) raised an interesting point with regard to the connection between bubbles and efficiency. He argued that, once one departs from the idealized world of perfectly functioning markets, where bubbles are detrimental to the well-functioning and efficiency of financial markets, bubbles may actually serve a beneficial purpose. He insisted that in some cases where the market is already biased due to structural imperfections like transaction costs, asymmetric information, etc., bubbles may be a device that helps to mitigate the market's structural problems. Nonetheless, despite these occasional reports of bubbles that are not undesirable, the present argument will continue on the much more common premise that most bubbles in market prices indicate an informational inefficiency which is potentially accompanied by negative repercussions for allocational and production efficiency.

2.2 Literature on Information and Derivative Markets

In Grossmann (1976), Grossman provided some of the most influential insights into the role of information in markets. He constructed a simple model of a market with a single risky asset and traders who can be either uninformed or become informed by incurring some cost. He reasoned that, in a perfect market with costly information, there must be noise so that agents can earn a return on their investment in information gathering. Otherwise the market will break down because it lacks both an equilibrium where agents earn a return on their information and one where agents do not gather information.

In reality, markets are not characterized by perfect information and noise is an ever-present fact in real-world financial exchanges. Recognizing this, in the 1970s finance research began asking the question of which markets are the first choice of traders who are in the possession of new or superior information. The results pointed away from spot, and toward derivatives exchanges. Several studies documented the propensity of information traders not to trade on their information in traditional stock markets. They are rather shown to take their business to options and futures markets, since these markets offer larger absolute returns with lower capital investment than the markets for the respective underlying. The major findings from these studies are summarized in the following paragraphs.

Manaster and Rendleman (1982) argued that in the long run, the instrument providing the greatest liquidity paired with the lowest trading costs and restrictions would be likely to play the predominant role in the market's determination of equilibrium stock prices. To support their conjecture that options are such an instrument, they argued that options entail relatively low trading costs compared to the underlying stocks. They are furthermore not subject to an uptick rule for the purpose of short-selling, may enable investors to reinvest the proceeds from such transactions, and come with lower margin requirements due to the higher leverage for a given investment amount.

In their empirical analysis, they calculated Black/Scholes-implied stock prices from option prices, using option price data from the CRSP tapes from April 26, 1973, to June 30, 1976, and weekly interest rate data from 91-day Treasury Bills. If options were priced according to the Black/Scholes model, these implied stock prices would be the option market's assessment of equilibrium stock values. They found that the difference between the implied and the observed stock prices (on day t) was positively related to returns on the stock on the following day ($t+1$). Furthermore, they could reject the hypothesis that the previous day's ($t-1$) implied stock prices contained no information concerning the following day's ($t+1$) return at the 1%-level. In their own words, "[...] there did appear to be evidence that closing option prices contained information that was not reflected in stock prices for a period of up to 24 h."[16]

Chern et al. (2008) used an event study approach of stock split announcements to compare stocks that were the underlying of an option (optioned stocks) to stocks that had no such accompanying option. They found a significantly greater anticipation of stock split announcements for optioned than for non-optioned stocks at the NYSE, AMEX and NASDAQ exchanges, conditional on there having been significant evidence of an anticipation of a particular stock split. They also reported a significantly smaller price reaction on the announcement day and on the following day for optioned NYSE and AMEX stocks. Taken together, this evidence supported their hypothesis that the announcement of a stock split conveys less new information in the case of optioned stocks than for non-optioned stocks, and that the former adjust more quickly to this information than the latter.

Figlewski and Webb (1993) echoed the arguments of Manaster and Rendleman (1982) in reasoning that option markets give traders who cannot or will not engage in short sales (e.g., due to transaction costs) an opportunity to sell short indirectly. They argued that the option market maker who is the counterparty of such a transaction will usually hedge by performing a short sale herself, subject to lower transaction costs and fewer constraints. Starting from this assumed mechanism, the authors conjectured that the existence of options should be positively related to the

[16] Manaster/Rendleman (1982), p. 1056.

average level of short interest.[17] They tested this hypothesis empirically using a sample of 342 stocks with uninterrupted data from 1969 to 1985 from the Standard & Poor's 500 index (S&P 500), taken from the CRSP tapes. The results show that relative short interest was significantly higher for stocks that had traded options than for those without, in each year of the sample.

Jennings and Starks (1986) examined quarterly earnings announcements from NYSE-listed stocks of the S&P 500 from June 15 to August 21, 1981, and from October 4 to December 31, 1982, to find what effect the trading of options on a stock had on the price impact of earnings announcements. They found that the prices of non-option companies took longer to adjust following earnings announcements than that of companies which were the underlying of option trading, supporting the notion that the latter were more efficient. Skinner (1990) arrived at similar results when he found that optioned stocks at the Chicago Board Options Exchange (CBOE) and the American Stock Exchange (AMEX) were being followed by a larger number of analysts than stocks without options written on them. He took that as an explanation for his second finding, namely that the stock price reaction upon the release of accounting earnings information for newly optioned stocks, as compared to levels prior to options being written on their shares, declined both in absolute terms and conditional on unexpected earnings, with significance at the 1%-level. Easley et al. (1998) showed that option volumes led stock price changes and carried information about future stock price changes, an interdependence that was later complemented by the results of Jayaraman et al. (2001). The latter reported that, for their sample period of 1986–1996, the CBOE led equity markets in terms of volume. Pan and Poteshman (2003) came to the same conclusion and reported that the effect was particularly evident for small stocks (which can generally be assumed to be less informationally efficient) and remained consistent at the annual level over a period of 12 years.

Lee and Yi (2001) found that informed traders preferred trading on the CBOE to trading on the NYSE, but not for all volumes. They calculated that large-volume informed trades were more frequent at the NYSE and argued that the reason for this observation may have been that large trades at the CBOE tended not to be anonymous, while they were more so at the NYSE. They argued that, since market makers at the CBOE could distinguish between informed and uninformed traders for larger orders, they increased the spread for informed traders, thus making the CBOE less attractive for such large informed orders. Furthermore, their results suggested that informed investors were attracted to options with lower option deltas, i.e., larger leverage.

Chakravarty et al. (2004) focused on a slightly different aspect of the topic and argued that informed insiders sometimes trade in option markets, a conjecture that they arrived at after reviewing insider trading convictions in option markets. They employed an approach first applied by Hasbrouck (1995), which allowed them to

[17] As a mechanism working in the opposite direction, they mention that the introduction of options may cause prior short sellers to switch their shorting activity to option markets, thus reducing short interest in the underlying. However, they believe this effect to be of inferior relevance, since short selling in stocks is relatively limited and because the hedging activities of the option counterparties would cancel out this effect to some degree.

measure directly the share of price discovery across 60 stocks listed at the NYSE that possessed options exclusively at the CBOE over a period from 1988 to 1992. With this method, they calculated implied stock prices from call option prices and compared them to actual prices in the stock market. The results showed that an average of between 17% and 18% of the price discovery occurred in the option market, with estimates for individual stocks ranging from close to 12–23% – numbers that they found to be significantly different from zero at the 1%-level. They also observed that the information share of out-of-the-money options seemed to be higher than for in- or at-the-money options, and that option market price discovery appeared to be an increasing function of volume – evidence that is consistent with informed traders who value both leverage and liquidity.

Schlag and Stoll (2005) broadened the research focus by analyzing both options and futures, again finding that (signed) options and futures volumes had a contemporaneous effect on the DAX price index in 1998. They investigated the source of price discovery in this market and found that futures traders possessed information about the index that was not reflected in the quotes, while the price effect of signed options volumes was largely temporary, which points to a liquidity (as opposed to an information-based) explanation. Interestingly, they also reported that signed futures volume led signed options volume. In an earlier article that focused only on futures markets, Cox (1976) developed a model to relate the effect of organized futures trading on spot market prices. Applying it to data from six different commodities over the years 1928–1971, he found evidence for more informed traders and a disappearance of spot price autocorrelation during periods of futures trading. Cao (1999) proposed a model which implied that the introduction of options caused an increase in the prices of the underlying asset and the market index, decreased the price response of the asset upon new public information, and increased the number of analysts following the underlying asset (consistent with Skinner (1990)). His empirical evidence backed up the predictions of the model, supporting his hypothesis that the installation of an options market induced investors to acquire more precise information, because it gave them additional opportunities to profit from trading on it.

Taken together, the evidence suggests relatively strongly that the presence of derivatives markets in general and option markets in particular tend to increase the efficiency and market quality in the market for the underlying stock. It were these results that formed part of the motivation for the experiments described in the following chapters.

2.3 Literature on Prediction Markets, Market Structure, and the Double Auction Mechanism

The phenomenon of prediction markets is a relatively new one, and even more so is the analysis of such markets by the economic literature. Nonetheless, the two decades since the introduction of prediction markets in 1990[18] have seen a

[18] Cp. Tziralis and Tatsiopoulos (2007), p. 75.

number of publications reporting on political stock markets, prediction markets used by companies to forecast future sales or project termination dates, and online betting sites. The steadily increasing number of studies dealing with this topic and the creation of the *Journal of Prediction Markets* by the University of Buckingham Press in 2007 indicate that the monotonicity of this increasing trend will not soon end. Because of their centrality to the research questions investigated in this book, the literature on prediction markets is reviewed below. The following paragraphs explore the reasons for individuals' participation in prediction markets, mention a study on a novel information aggregation procedure, and provide evidence on the performance of prediction markets with abstract underlyings. They furthermore briefly discuss markets in the fields of finance, sports and politics.

In the first formal theoretical study of prediction markets, Forsythe et al. (1992) explored why individuals would spend time trading in such a market. Specifically, they listed five motivations for traders to participate in a political stock market experiment, which were (1) entertainment, (2) expected differences in information (confidence in their knowledge about the political event relative to other traders), (3) expected differences in information-processing ability (confidence in their ability to interpret news relative to other traders), (4) expected differences in their talents as traders, and (5) risk-seeking behavior. Forsythe et al. expected these differences to attract a diverse group of experimental subjects and were able to confirm this belief when analyzing actual political stock market participants' demographic characteristics, political and ideological preferences, investments, and earnings.

In the context of prediction markets, another issue of considerable practical importance (originally identified by Manski (2004)) is under which conditions prediction market prices reflect the true aggregate beliefs of the individual traders. To explore this issue, Wolfers and Zitzewitz (2006) proposed two simple models based on a log utility function, which lead to an equilibrium price in the market that is equal to the mean belief of traders. They then went on to relax some of the simplifying assumptions, showing that the dual symmetry assumptions of (1) demand being a function of the difference between beliefs and market prices which is symmetric around zero, and (2) a symmetric distribution of beliefs, lead to the same result (i.e., equilibrium prices being equal to the mean belief) without the need for log utility. They also found that if wealth and beliefs are not orthogonal, the equilibrium price turns out to be a wealth-weighted average of individual beliefs. Once the dual symmetry assumptions were also dropped, the possibility was raised that prices deviate from mean beliefs, but the authors argued that these deviations remain small under most reasonable specifications of utility and distributions of beliefs.

In a third theoretical inquiry into the properties of markets as information gathering tools, Plott (2000) set out by questioning whether it is at all possible that a market aggregates and processes the immense number of simultaneous equations and inequations expressing investors' beliefs, preferences, and differential information. In answer to this question, he then reasoned that this process is simplified by investors themselves, since each investor reaches his opinion of the "correct" price not only by considering the information she herself is privy to, but also forms expectations of the information others possess and of the beliefs others will form. Switching from theoretical to empirical argumentation, Plott then described a laboratory experiment in which he showed that an experimental market was indeed capable of extracting a larger set of information from the transactions of experimental subjects, each of whom had gotten only a small bit of the full information set regarding the value of an abstract underlying asset. In a similar vein, Wolfers and Zitzewitz (2004) also provided encouraging testimony of the ability of prediction markets to forecast uncertain future events. They found that "[...] simple market designs can elicit expected means or probabilities, more complex markets can elicit variances, and contingent markets can be used to elicit the market's expectations of covariances and correlations [. . .]"[19]

Berg et al. (2003) used the Iowa Electronic Market's prediction of the outcomes of the 1988, 1992, 1996 and 2000 U.S. presidential elections to provide the first study of the long-run predictive power of forecasting markets, finding that their markets gave accurate forecasts at both short and long horizons (single day vs. weeks and months). They then compared the predictions of the Iowa Electronic Market to the forecasts of various polling organizations, reporting that the latter were being outperformed by the former.[20] In another study on the predictive power of prediction markets, Tetlock (2004) used data from tradesports.com, an online market which at that time allowed wagers on both sports events and financial market data. He showed that financial prediction markets can be surprisingly efficient with relatively low numbers of market participants. His study also documented that results from sports wagering markets may not be replicable in economic prediction markets, since inefficiencies in the former segment of his sample did not reappear in the latter.

In contrast to the studies discussed so far, Ortner (1996) reported results from prediction markets run on election outcomes in Austria, where markets showed clear signs of manipulation and did not reliably provide forecasts of higher quality than polling organizations. Rather, the market's results in his experiment had been deliberately and successfully manipulated by a minority of traders to deviate from the market's earlier consensus opinion, at the same time influencing the prices of related markets. Chen et al. (2003a) also deviated from the bulk of the prediction market literature, albeit in an entirely different way. While most studies reported on markets employing standard double auctions, in their experiment they performed a nonlinear aggregation of individuals' predictions based on said individuals' skills and risk attitudes, as determined in previous prediction rounds in the same market.

[19] Wolfers and Zitzewitz (2004), p. 124.

[20] In section I of their paper, they also gave a good overview of online prediction markets in existence at the time of their publication (see Berg et al. (2003), pp. 2–3).

The results from such a "weighted" prediction outperformed both the simple market and the best of the individuals.

Overall, the diverse topics of studies on prediction markets and their heterogeneous findings underline the novelty of the field. While not specifically focusing on prediction markets, this study nonetheless offers new evidence on markets' ability to process information and harmonize expectations.

2.4 Literature on Experiments in Economics[21]

Economists began analyzing the special properties and functioning principles of market-based exchange in the eighteenth and beginning nineteenth century, starting with the work of Adam Smith[22] and Antoine Augustine Cournot. While the use of laboratory experiments in economics dates back to about the same timeframe (cp. Bernoulli (1738), as argued in Roth (1995)), the beginning of its widespread adoption by a sizable number of economists took place no earlier than in the twentieth century.

Generally, experimentation in economics can be segregated into three different research directions – those of game theoretic experiments, individual decision-making experiments, and market experiments.[23] The latter, which is the line of research the present study fits into, had its origin in the work of Chamberlin (1948). Chamberlin performed a laboratory market experiment by assigning reservation prices to a number of student subjects and allowing them to roam around the classroom with the goal of finding partners to trade with. He reported finding transaction volume in excess of the equilibrium quantity in 42 out of 46 markets and mean prices below the equilibrium price in 39 cases. Due to the substantial deviation of these results from theoretical predictions, Chamberlin dismissed them after one publication and discontinued his experimental research. While Chamberlin had thus laid the groundwork with his initial experiments, it was his student Vernon Smith (1962, 1964) who made experimentation the center of his life's research effort. It is a sign of the importance experimentation has since gained in economics, that in 2002 the Royal Swedish Academy of Sciences awarded him with the Sveriges Riksbank Prize in Economic Sciences in Memory of Alfred Nobel, "for having established laboratory experiments as a tool in empirical economic analysis, especially in the study of alternative market mechanisms."[24] Before that, the award of the prize to Maurice Allais in 1988 and to Reinhard

[21] Cp. Davis and Holt (1993) and Roth (1995).

[22] Cp. Smith (1843).

[23] There are also experiments like those of Williams and Walker (1993), which serve no research question but are conducted in university classes to introduce student subjects to topics from the field of microeconomics.

[24] Royal Swedish Academy of Sciences (2002).

Selten in 1994 could be considered indirect signs of recognition of the importance of experimentation, which had featured in a prominent role in Allais' tests of game theoretic concepts and in Selten's work on individual behavior.[25]

Compared to traditional empirical studies, experimentation under controlled conditions has the advantage that single parameters may be varied while keeping all other conditions constant, thereby allowing for the isolation of the effect of variations in single variables. In natural data, tests of market propositions are always tests of the joint hypotheses of the primary hypotheses to be tested and the auxiliary hypotheses regarding the general market situation, equilibrium, agents, and a plethora of other circumstances. Any result, be it supportive or contradictory, may under these circumstances be caused either by mechanics implied in the primary hypotheses, or be due to erroneous auxiliary hypotheses. Conducting controlled experiments alleviates this problem by allowing the experimenter to reduce the number of auxiliary hypotheses. Experimentation also enables the researcher to obtain repeat observations under identical conditions, an important prerequisite for the analysis of the robustness of results. This advantage is all the more important since empirical data – if available – is usually expensive, while at the same time often lacking in accuracy.

Nonetheless, experimental economics has been subject to strong criticism over the years. One point of criticism is that a majority of economic experiments employs student subjects, raising the concern that this group is not representative of agents in real economic contexts. The results of studies testing this proposition somewhat invalidate this argument; they are reported in Sect. 2.4.4.6. Another concern is that the simplification of markets, the environment and the sets of possible actions in laboratories yield results that are not meaningful when applied to real-world markets. This is a valid point which must, however, also be applied to theoretical research and model building; just as in experimental research, simplification is a necessary component of this strand of research. Besides, experimental studies hold the possibility to probe the impact of these simplifications, by varying individual parameters and measuring their impact on the results. Laboratory markets have also been criticized as not being "real," an argument that Plott (1982) countered by pointing out that, in the context of experimental markets, the same principles of economics apply as elsewhere. As he put it, "Real people pursue real profits within the context of real rules."[26] He noted that the simplicity of laboratory markets should not be confused with the question about their reality as markets.

Smith (1994) listed a number of reasons from the literature as to why economists conduct experiments, among them the wish to test a theory or explore the reason for its failure, the observation of empirical regularities as a basis for a new theory, the comparison of environments and institutions, and the evaluation of policy proposals and test of institutional design. The present book set out to do the last, i.e., test the

[25] Cp. Haase (2006), p. 166–167.

[26] Plott (1982), p. 1520.

impact of digital option trading on spot market efficiency. As mentioned in the introduction, the observation of empirical regularities then led to the formulation of a new hypothesis. This work thus is a good illustration of one of the points Smith (1994) made, namely that experimentation has many dimensions and can shed light on topics of scientific research in a variety of ways.

In their book surveying the whole discipline of experimental economics, Davis and Holt finally drew the following conclusion regarding the value of experimentation as a research methodology in economics:[27]

> "Overall, the advantages of experimentation are decisive. Experimental methods, however, complement rather than substitute for other empirical techniques. Moreover, in some contexts we can hope to learn relatively little from experimentation."

One can summarize the above deliberations by noting that the experimental method is one of a number of instruments in the economist's toolbox. Its value depends on the research question under examination, yet it is able to address issues that are hard – if not impossible – to tackle with alternative approaches. In the case of the research question addressed by this study, its advantages by far outweighed its shortcomings, a point that will become clearer in the discussion of the results in Chap. 4.

2.4.1 Expectations and Equilibrium Models in Experimental Asset Markets

Models are to be used, not believed.
Henri Theil (1971)

The question of efficiency and inefficiency in any market, both inside and outside of the laboratory, is intricately intertwined with that of the formation of expectations by the market participants. The topic of expectation formation has been a staple of economics research for a number of decades, but received additional momentum with the advent of behavioral finance and the increasing influx of results from psychology and biology into the economic sciences. For this reason, this literature is reviewed in this section. As will become clearer during the discussion of the results in Chap. 4, the process and mechanics of expectation formation are of central importance for this work.

2.4.1.1 Prior Information Equilibrium

Plott and Sunder (1988) defined a *prior information equilibrium* (also referred to as a naive price equilibrium in Forsythe et al. (1982)), as an equilibrium following from the actions of agents which consider only their private information

[27] Davis and Holt (1993), p. 18.

in investment decisions. In other words, in such an equilibrium, individuals evaluate prices based solely on their own information – ignoring the possibility that market prices, by aggregating information from other traders, also contain information. They are assumed to apply Bayes' law to determine the likelihood of a state of nature given their private (prior) information. After having done so, they maximize their utility dependent on that likelihood, but do not take into account market prices and possible speculation potential depending on the actions of other market participants.

The prior information equilibrium does not play a major role in the experimental literature, but is sometimes used as a somewhat extreme bound on subjects' behavior. By benchmarking experimental results against the expectation formation mechanism implied in this equilibrium model, strong deviations from the predictions of rational expectations theory can sometimes be better illustrated, or statements can be made regarding the (lack of) plausibility of results (for an example, see e.g., Plott and Sunder (1982)).

2.4.1.2 Rational Expectations Equilibrium

Our fundamental view is that the experimentalist has as much to learn from experimental subjects about subjective rationality, as human decision makers have to learn from the models that we call "rational."

Vernon L. Smith and James M. Walker (1993b)

Smith et al. (1988) distinguished between two definitions of rational expectations. They quoted the more common Muth (1961) definition that rational expectations for the same information set tend to be distributed about the prediction of the theory,[28] as well as the earlier and less restrictive Nash (1950) definition, that for expectations to be rational, they should be realizable.[29] They interpreted the difference to be that rational expectations according to Nash (1950) need to be sustained or reinforced by outcomes, while rational expectations as defined by Muth (1961) are implied to be sustained by outcomes that in turn support theoretical predictions. Specifically, Muth wrote that "the expectations of firms (or, more generally, the subjective probability distribution of outcomes) tend to be distributed, for the same information set, about the prediction of the theory (or the 'objective' probability distributions of outcomes)." In short, the rational expectations hypothesis states that the expected price is an unbiased predictor of the actual price. Muth qualified this statement by saying that it holds true only in the aggregate.[30]

A theory of rational choice that is considerably more realistic, albeit much harder to operationalize than the above concepts, is that described by Simon

[28] Cp. Smith et al. (1988), pp. 1136–1137, and Muth (1961), p. 316.

[29] Cp. Smith et al. (1988), p. 1137 and Nash (1950), p. 158.

[30] Cp. Muth (1961), p. 333.

(1955). He drew a picture of individual behavior that is characterized by bounded rationality and a variety of coping strategies. Agents in this model curtail the set of all possible actions to derive a subset of actions they take into consideration in their decision-making process. Furthermore, they do not optimize the expected outcome over the available alternative space of actions, but employ a strategy of satisficing, i.e., choosing an action that leads to expected outcomes which satisfy some subjectively set minimum acceptable level, as opposed to providing the maximum possible benefits.

Despite its better fit with reality (and with a large portion of experimental results), Simon's (1955) notion of rational choice has not been widely adopted in the experimental literature.[31] This is possibly due to the difficulty of operationalizing its predictions in real applications. Muth's (1961) definition, which is the concept most often employed in the literature, often fails in describing the actual behavior of subjects encountered in experimental and empirical studies. Yet the beauty of its predictions is that they constitute a natural upper bound on the possible extent to which individuals *can* adhere to models based on the assumption of homines oeconomici. One of the stated objectives of many economists is the discovery of a market structure that does the best possible job of processing information, so that asset prices correctly and completely reflect the available information set. Testing the performance of any given market system by comparing its outcomes to the predictions of the theory of rational choice as formulated by Simon (1955) might yield a broad congruence between prediction and outcomes, yet it would not further the objective of finding a market structure that optimizes information processing and price efficiency in line with economic theory. This is something that a comparison with the predictions of a model of behavior following Muth (1961) and a program of minimizing the deviations of actual outcomes from the results predicted by his concept of rationality would accomplish.

Note that it is somewhat dangerous to use the word "rational" in the context of such discussions. While it is tempting (and common practice) to refer to individuals resembling the theoretical concept of the homo oeconomicus as being rational, this is correct only when abstracting from e.g., the cost of thinking. When Smith (1985) talked about the modification of standard theories by introducing elements of the subjective cost of transacting, he (correctly) referred to this as "imbedding standard theories in larger (and more '*rational*') frameworks" [italics added for emphasis]. In reality, individuals who take into consideration the cost of finding an optimal solution (in terms of cognitive effort, time dedicated to search, etc.) should be referred to as more rational than their compatriot homines oeconomici, who pursue optimality regardless of the cost of this pursuit. Nonetheless, unless otherwise noted, this study will follow the conventional practice of equating rationality in economic decision-making with adherence to the theoretical model of the homo oeconomicus, which coincides with Muth's (1961) definition of rational expectations.

[31] It has found more adherents in the literature on behavioral finance and decision-making.

Another important observation regarding this topic is the difference between the meaning of rational expectations in the market efficiency literature and in the experimental literature. While in the former, market efficiency is a characteristic of a given platform of exchange, in the latter it is the result of a process. Plott and Sunder (1988) summarized this when they wrote:[32] "Rational expectations can be seen either as a static theory of markets (e.g., in the efficient market literature in finance) or as an end-point of a dynamic path of adjustment." Most experimental evidence indicates that expectations are adaptive and that rationality may take some time to settle in (if it does at all). A study that nicely illustrates the importance of the adaptation of expectations (and its speed) was Arthur et al. (1996). The authors proposed a model of rational, heterogeneous agents who endogenously form expectations about market prices, which are subject to influence from their own decisions. In doing so, these agents assign a positive probability to the existence of irrational agents – in other words, the rationality of all agents is not common knowledge. With this setup, the authors wished to explore the question of whether such a market leads to an evolution toward homogeneous (rational expectations) beliefs or whether it exhibits more varied behavioral patterns which could explain some of the seemingly irrationality-motivated phenomena in real-world markets. Simulating markets with the characteristics described above, they found that both outcomes were possible and robust over certain subsets of the parameter space. If they parameterized their agents in a way that had them adapting their forecasts unrealistically slowly, the market converged to a rational expectations equilibrium. In parameterizations where forecasts were adapted at a more realistic rate, behaviors in the market did not converge and pseudo-psychological effects like bubbles and profitability of technical trading rules could be observed. In this latter design, they also found persistence in volatility and trading volume, as well as GARCH effects. Williams (1987) arrived at a similar verdict after showing experimentally that subjects are not Muthian rational when forecasting experimental double auction market prices. Forecasts in his study turned out to be biased with regard to the mean price, and to display significant first-order serial correlation. He concluded that an adaptive expectations model describes the experimental regularities better, a finding that was arrived at also in a large number of other experimental studies, including Smith et al. (1988) and this present study.

An interesting twist on the topic of expectation formation and rationality was discovered by Frédéric Koessler, Charles Noussair and Anthony Ziegelmeyer. In Koessler et al. (2005) they documented that the elicitation of beliefs from experimental subjects moved their choices in a parimutuel betting market closer to those predicted by a rational expectations model.[33] It also increased the amount of information aggregated in prices. They found that – without requiring subjects to state their expectations – public information was being overweighted relative to each subject's private information. Once subjects were asked to submit their beliefs

[32] Plott and Sunder (1988), p. 1104, footnote 6.

[33] In a parimutuel betting system, all bets are pooled and later shared among the winning tickets.

regarding future outcomes, they started placing more weight on their private information relative to public information, leading to a more efficient aggregation of the existing information into prices.[34] Furthermore, in cases where subjects had erroneous private information, it also induced them to more often follow public information that was (correctly) in conflict with their private information. Such a phenomenon is called an *information cascade* in the literature. Alevy et al. (2007) described it as follows:[35]

> "Information cascades arise when individuals rationally choose identical actions despite having different private information."

In the same article, Alevy et al. also pointed out that this is a phenomenon that is distinct from herding, as the latter does not necessarily involve rational individuals, but can be caused by preferences for conformity, social sanctions or lower necessary cognitive effort.

2.4.1.3 Perfect Foresight Equilibrium

As the final theoretical model in their article, Forsythe et al. (1982) listed the perfect foresight equilibrium (which in the case of their experiments equaled the rational expectations equilibrium), also referred to as a fully revealing rational expectations equilibrium in Plott and Sunder (1988). In this theoretical model, agents behave as if they had the perfectly forecasted theoretical equilibrium price at their disposal. In other words this is the rational price a homo oeconomicus-type investor would arrive at were he in possession of full information. In their experimental work, Forsythe et al. (1982) then found that the rational expectations equilibrium (i.e., the perfect foresight equilibrium) was an excellent predictor of the performance of their simple markets and that replication was both a necessary and sufficient condition for the applicability of the perfect foresight model. They reported that none of their five experimental markets converged in the first period, while all of them converged after replication.[36] Forsythe et al. (1984) similarly showed that the perfect foresight model was a good predictor of the last several years in experiments with spot- and futures markets, whereas in their sequential markets it was a good predictor only of the final year.

In the latter article, Forsythe et al. also analytically compared whether final-year allocations were more accurately predicted by the perfect foresight model than by the prior information model. They found that the perfect foresight equilibrium model was a good predictor of allocations in late years (always better in years six and seven), while the prior information model did better in the early years (nearly

---

[34] Cp. Koessler et al. (2005), p. 14.

[35] Alevy et al. (2007), p. 151.

[36] Cp. Forsythe et al. (1982), p. 560: "The appropriate model may have the markets converging to a temporary (naive) equilibrium *first* and then adjusting to the perfect foresight equilibrium after "sufficient" information has accumulated [...]"

always better in years one and two) of their experimental markets. This again matched the observations reported in their earlier study. Camerer and Weigelt (1991) reported similar results in their article on the occurrence of information mirages, which is briefly discussed in Sect. 2.4.2.2. Both observations support the view of rationality as the result of a learning process within subjects.

2.4.1.4 Maximin Equilibrium

In addition to the Forsythe et al. (1982) models, Plott and Sunder (1988) described the maximin model, which is characterized by agents who act only on certain payoffs. In the maximin framework, the investors with the maximum (across all traders) of minimum (across all states) dividends will purchase the security at a price equaling their minimum dividend. Note that this equilibrium does not apply to experiments like Smith et al. (1988), where all investors face the same reservation cost and value for one unit of the experimental asset.

2.4.1.5 No-Trade Equilibrium

2.4.1.5.1 No-Trade Equilibrium in a Stock Market

A final possible equilibrium in many experimental markets is one where no trade takes place. This equilibrium is of particular interest for the discussion of the experimental results reported in later chapters, because it is frequently conjectured to be the "rational" equilibrium for Smith et al. (1988)-type asset markets. However, such an equilibrium requires the following five relatively restrictive conditions to hold:[37]

Condition 1: The initial cash and asset allocation is Pareto optimal.
Condition 2: All subjects are rational maximizers of expected utility.
Condition 3: Condition 2 is common knowledge.
Condition 4: Subjects derive utility only from final payoffs, not from the process of trading itself.
Condition 5: There are no cognitive or transaction costs to trading.

If the first condition is violated, trade is nonetheless limited to Pareto-improving transactions and will not display patterns where a subject for example first buys an asset and then sells it again, or vice versa. Once a Pareto optimal situation has been reached in such a market, trading once again ceases.[38] A violation of condition

[37] The author wishes to thank Erik Theissen for suggesting conditions one to three.

[38] The argument assumes that over the time of the laboratory experiment, subjects' preferences are constant and that changes in subjects' wealth due to the receipt of dividends over the experimental periods are insufficient to change their optimal portfolio sufficiently to induce subjects to develop the wish to rebalance their portfolios.

three has in turn been proposed as an explanation for many of the inefficiencies (in particular price bubbles) observed in experimental asset markets. As Lei et al. (2001) showed and as this book will also suggest, this explanation is not sufficient to explain the observations. The reason behind this is that in some designs common knowledge is irrelevant for market efficiency, yet inefficiencies are still observed. However, a violation of condition two can explain the results found by Lei et al. (2001) and is also consistent with the literature on bounded rationality. In a market with less than perfectly rational subjects, trade is possible even if the initial allocation is Pareto efficient.

Furthermore, trade is also possible in any market where subjects derive utility directly from the act of trading. Such a mechanism is suggested by the Active Participation Hypothesis proposed by Lei et al. (2001). It implies that subjects in experimental markets trade because they feel that they are supposed to trade, even if it does not increase their expected utility from the final future payoff. It is also consistent with Williams' observation that subjects in his experiment were so fascinated by the electronic trading mechanism that they traded significantly more than expected.[39] Finally, the fifth condition ensures that no considerations other than those of final payoffs bias subjects' actions.

Note that, if the five conditions above hold, there will be no trade, but there may be quotes (i.e., limit orders). If the initial allocation is Pareto optimal, but this is not common knowledge, even rational individuals (who do not know that they are in a situation of pareto optimality) may try to improve their situation by offering trades. However, due to Condition 1, no other individual will want to take the opposite side of any such quote. On the other hand, consider what happens in a market where the following condition is introduced:

Condition 6: Condition 1 is common knowledge.

In the case where Conditions 1 through 6 hold, the market will not only exhibit a no-trade equilibrium, but there will not even be any quotes, since every trader knows that no other trader will transact with her.

2.4.1.5.2 No-Trade Equilibrium in a Digital Option Market

Digital options are characterized by a trinary payoff structure that makes them unsuitable for hedging purposes.[40] To be suitable for hedging, an instrument needs to have a payoff structure in which the marginal payoff – at least over some parameter interval – runs opposite (or parallel) to that of the asset to be hedged.

[39] Williams (1980), p. 245.

[40] The payoff structure of the digital options employed in the empirical part of this study will be described in detail in Sect. 3.3. In short, a digital option pays a fixed amount to the winning party, pays nothing to the losing party, and splits the payoff equally in the case where the price of the underlying equals the option strike price at maturity (i.e., when the option is at the money at maturity).

This is not the case for digital options, which makes them primarily a vehicle for speculation. Because of this reason, it cannot be argued that subjects use digital options to improve their risk exposure, but only to improve their cash position. It must be assumed that they contract for digital options only if their subjectively perceived expected value from the digital option investment is positive.[41] Following this argument, a no-trade equilibrium in the digital option market relies on the following conditions:[42]

Condition 1: All subjects have homogeneous expectations.
Condition 2: All subjects are rational maximizers of expected utility.
Condition 3: Subjects derive utility only from final payoffs, not from the process of trading itself.
Condition 4: There are no cognitive or transaction costs to trading.

The interpretation of violations of Conditions 2 through 4 is analogous to the section above. Condition 1 is new in that the allocation of cash and assets is irrelevant when regarding digital options, yet the form of expectations about the future price of the underlying is critical. If Condition 1 is violated, investors will trade on their asymmetric information or on their heterogeneous interpretation of symmetric information (i.e., their heterogeneous expectations based on symmetric information, due to heterogeneous beliefs). Furthermore, consider the following condition:

Condition 5: Conditions 1 and 2 are common knowledge.

Condition 5 can be employed to make a similar argument as Condition 6 in the section on the stock market above. If it holds, then – in addition to there being no trade in such a market – no market participant will even post digital option offers (i.e., limit orders in the digital option market), since everybody is aware that no other trader would enter into an option contract that the first trader would consider favorable.

2.4.2 The Role of Experience in Experimental Asset Markets

The twin issues of learning and experience play a prominent role in any science investigating the actions and behavior of humans, regardless of the context. For the discipline of economics, Friedman et al. (1983) distinguished between

[41]This argument assumes that subjects are not risk-loving and is developed in more detail in Sect. 3.3.

[42]This analysis assumes that subjects cannot influence the future price of the underlying. If, as in the experimental market described later in this text, the same individuals trade both in the digital option market and in the market for the underlying, then a no-trade equilibrium in the option market also requires Condition 3 from the analysis of the stock market above – the condition that rationality be common knowledge.

three types of experience relevant in experimental asset markets, which would also lend themselves to generalization to other sciences employing systematic experimentation:[43]

> "In a real-time trading process such as that of our experiments, equilibrium can be achieved only as agents learn about their opportunities for gain through trade. In our experiments this learning can take place within each period as traders observe bids, offers and transactions (intra-period learning) – across periods and market years as traders observe trends in prices and the outcomes of their activities (inter-period learning), and across experiments as traders gain a better idea of what information is relevant and refine their strategies (experience)."

These three terms – intra-period learning, inter-period learning and experience – will be adopted for the purposes of this text. However, since no study reviewed for this book analyzed intra-period learning, the first category will be disregarded in the following literature overview.

2.4.2.1 Inter-Period Learning

The article of Forsythe et al. (1982) was already mentioned in Sect. 2.4.1 on equilibria, but shall be mentioned here again because of the relevance of its results for the topic of inter-period learning. Forsythe et al. (1982) found for their design that replication (i.e., the repetition of experimental runs with the same treatment, or "intra-treatment" experience) is both a necessary and sufficient condition for convergence to the price predicted by the perfect foresight model. Friedman et al. (1983) reported that in each of their four experiments (with the exception of a single period in one experiment) profits were generally higher in later market years. Friedman et al. (1984) also reported that their markets converged over time toward informationally efficient equilibria. In their experiments, this finding was robust to the presence or absence of futures markets and to that of uncertainty regarding the future state of nature.

Smith et al. (1988) reported on three of their experiments which in the first three periods seemed to converge to, and from then on closely followed the path of expected dividend value. Even in these experiments they found support for the conclusion that the rational expectations model of asset pricing can be confirmed only as an equilibrium concept underlying an adaptive price adjustment process. This is in conflict with Fama's concept of efficient capital markets, which requires that "security prices *at any time* 'fully reflect' all available information"[44] [italics added for emphasis].

In an experiment which forms a connection to Sect. 2.2 on the role of information in experimental markets, Camerer and Weigelt (1991) ran an experimental asset market where the subjects faced uncertainty about the presence of informed

---

[43] Friedman et al. (1983), p. 130.

[44] Fama (1970), p. 383.

traders. Their treatments ran for between 15 and 21 periods and in the majority of their experiments the probability of insiders being present was 0.5 in each period. Their findings suggested that subjects sometimes wrongly interpreted price patterns as stemming from insider trades, which then caused them to trade on noise as if it were information. They dubbed this phenomenon an "information mirage." Analyzing their time series data, they found that no mirages occurred in later periods and concluded that traders learned to distinguish between insider and non-insider periods using non-price information (i.e., the speed at which trading took place).

Providing some more detailed evidence on the equilibrating process or the process of expectations formation, Peterson (1993) reported that inexperienced subjects submitted forecasts which were frequently biased and inconsistent with the rational expectations hypothesis. However, these subjects altered their learning model more often than experienced subjects, and usually in the direction of rational expectations. This suggests an asymptotic learning process with a steep learning curve for inexperienced individuals, which flattens as they gain experience and approach the forecasting model implied by the rational expectations hypothesis.

2.4.2.2 Experience

The literature knows mixed results regarding the question of which impact experience has on the results of experimental studies. Most articles report that experience increases efficiency and rationality, reduces the variance of subjects' actions and – in experiments where this is possible – increases subjects' profits. Nevertheless, expert subjects (which are frequently assumed to be experienced) do not consistently outperform inexperienced students in terms of rational behavior. As in the above sections, the below paragraphs will review the relevant literature on this topic.

In an early computerized experiment, reported in Williams (1980), inexperienced subjects failed to achieve as rapid a convergence to efficient prices as documented in earlier, oral double auction studies. While this result is not particularly spectacular in itself, the reason that Arlington Williams believed to have been the cause for it may seem amusing from today's point of view. As his following statement suggests, the result may not only have been due to the complexity of the economic task, but may rather have been caused to a considerable degree by his subjects' unfamiliarity with the computer interface:[45]

> "In conducting the first series of [computerized double auction] experiments it became apparent that the ocular-motor skills required to function well in [computerized double auction] markets generally developed after a few periods of trading but seemed to totally elude some people."[46]

---

[45] Williams (1980), pp. 251–252.

[46] This example nicely illustrates the role experimental institutions play for the results and should serve as a cautionary tale for inexperienced experimenters. Note that the computer interface used for the experiments reported in later chapters was tested and adapted extensively prior to its first use in a live experimental session.

Nonetheless, when repeating the experiments with experienced subjects (who had shown themselves adept at grasping the computerized double auction mechanism) Williams found that the price convergence was faster and the market generally more efficient than in experiments with inexperienced subjects.

Similar findings – at least with regard to efficiency – were provided by Friedman et al. (1983), an article reporting on four markets: Two with inexperienced subjects, the other two with experienced subjects, and each with three periods per market year.[47] Friedman and his co-authors found that the dispersion of the transaction prices of inexperienced traders was consistently larger than that for experienced traders in their experiments, and that the latter had consistently smaller coefficients of variation. The authors interpreted this to mean that the experienced traders held probability beliefs with greater precision than the inexperienced subjects and would not accept bids or offers too far removed from the expected equilibrium price. They could also solidly reject the hypothesis that the mean transacted period B spot price converged to the perfect foresight equilibrium price for inexperienced traders, while equally firmly accepting the hypothesis for the experienced traders. Furthermore, aggregate profits in the experiments with experienced traders were all higher than those of the inexperienced traders.

In their seminal 1988 article, Smith, Suchanek, and Williams also dedicated considerable attention to the role of subject experience. Prior to their actual experimental sessions, they ran pilot experiments of their asset markets and found that subjects with no previous double auction experience of any kind (provided with relatively little information) produced prices deviating widely from the expected future dividend values. Thus, for their non-pilot experiments they used only once-experienced subjects and provided them with more information. After repeatedly observing price bubbles in markets with experienced subjects, they then conjectured that the bubbles with first-time traders were due to their inexperience, while experienced traders produced bubbles because they had gained their prior experience in a market that had similarly exhibited a bubble. To control for this possibility, they let inexperienced traders gain their first double auction experience in a market that was reinitialized after each period, so that no capital gains or losses were possible across trading periods.[48] However, these newly experienced traders, who had no prior experience of a market that had exhibited the bubble phenomenon, nonetheless produced bubbles when allowed to trade in Smith et al.'s (1988) baseline markets without reinitialization. The three authors also conducted a market experiment populated only with twice experienced subjects who had been among the top earning traders in previous rounds. The resulting bubble was similar to those observed in earlier experiments. Finally, they found that if a group of experienced

[47] A slightly more detailed account of the period design can be found in Sect. 2.4.4.2 on futures markets.

[48] They reported that the single period markets did also not exhibit any within-period price bubbles.

traders participated in two (additional) rounds, they no longer produced bubbles in the second.

Van Boening et al. (1993) similarly reported on a connection between experience and efficiency. They let subjects participate in a series of three markets, in order to collect data from one market design with the same group of subjects having first no experience in experimental asset markets, then being once-experienced and finally being twice-experienced. This design had already been used by Smith et al. (1988) and was also employed for this study.[49,50] Despite alterations in the trading institution (they used a closed-book call auction) and the dividend distribution (described in Sect. 2.4.4.1) they found that the only parameter that led to a decrease in price deviations from intrinsic value was an increase in subject experience. In a slightly later article, Porter and Smith (1995) also reported on the importance of experience, stating that their empirical evidence showed that inexperienced subjects tended to produce bubbles and crashes relative to a declining expected dividend value, while once-experienced subjects produced a less pronounced pattern of the same form that then practically disappeared for twice-experienced subjects.[51]

Oechssler et al. (2007) ran experiments of a somewhat different design and discovered a rare counterexample to the pattern of experience increasing price efficiency. Their subjects could trade five different assets, and in each session, one of these assets paid an extra dividend. The authors found that in treatments where the asset that carried this extra dividend changed from session to session, experience (up to two replications) did not lead to a reduction in the frequency of bubbles.

Dufwenberg et al. (2005) departed from the norm of having either only inexperienced or only experienced subjects in an experiment. They populated their markets with one third (two thirds) inexperienced traders and two thirds (one third) traders thrice experienced in a market similar to that employed in Smith et al. (1988). They found that in both treatments (one and two thirds experienced subjects) bubble-and-crash patterns were greatly reduced compared to the baseline case. Regarding a similar question, Ackert and Church (2001) reported no significant difference in price deviations from fundamental values between markets populated solely by experienced business or arts and sciences students and markets

---

[49] The only difference in this regard between Van Boening et al. (1993) and Smith et al. (1988) vs. this study is that, for procedural reasons, repetitions were conducted on the same day for this work, while the earlier articles invited subjects for experimental runs on different days. This topic will be elaborated upon in Sect. 3.1.

[50] Such a design is referred to as a within-subjects design in experimentation, pointing to the fact that differences in results from one round to the next – barring any changes in the experimental environment – must be due to changes within subjects, whereas in a between-subjects design, different results may be caused by different experimental subjects. Where possible, experimenters tend to prefer within-subjects designs, because they offer less possibility for noise to influence results.

[51] Cp. Porter and Smith (1995), p. 509.

made up of 43–50% inexperienced subjects and 50–57% (mostly twice) experienced subjects. They concluded that in the mixed markets, the subset of experienced traders was largely responsible for price-setting.[52]

Using a different modification in the subjects variable, Leitner and Schmidt (2006) wrote that, in forecasting tasks, expert subjects generally perform well in domains with static stimuli (e.g., weather forecasts), whereas they perform poorly in environments of dynamic stimuli and human behavior, such as financial markets. In their empirical study, they compared expert subject forecasts of the EUR/USD exchange rate from January 1999 to March 2003 with the forecasts of inexperienced students. To further enrich their tests, the students were provided with no other information than the realizations of the time series; they were not even told what kind of time series it was they were seeing and forecasting. The comparison of the two forecasts was based on three measures of efficiency: unbiasedness, absence of serial correlation in the forecast errors and efficient use of information. The results showed that all forecasts (from students and experts) of the horizons of 3 and 6 months exhibited significant correlation of forecast errors and made inefficient use of information at time lags of 1 and 2. The only efficient forecast (according to all three criteria named above) was that of the student subjects for 1 month ahead. More generally, the experts in their study seemed to expect trend reversals, while the students predicted short-term continuation of trends, with reversals in the long run, which corresponds to the short-term momentum and long-term reversal results for stock markets in the efficient markets literature, as mentioned in Sect. 2.1.2. Furthermore, the experts exhibited a bias toward fundamentals (i.e., purchasing power parity) in their forecasts, whereas the students' predictors did not. Overall they concluded that the experts' forecasts were significantly worse than the naïve student forecasts, a result which they could not attribute to a common failure in human decision-making.

2.4.3 The Baseline Experimental Market and its Extensions

Chamberlin (1948), and later Smith in his early work on double auctions, induced differing values of the experimental asset by assigning differing reservation costs and values to subjects. Later work, starting with Smith et al. (1988), assigned the same value to each unit of the asset, regardless of which trader ended up owning it at the end of the experiment. To the surprise of the experimenters, subjects generated trading volumes that far exceeded all bounds that could have been explained by differences in endowments or risk attitudes. The reason their article sparked a large number of additional studies, however, was the observation of price bubbles and crashes in their setting. The following sections present the original Smith et al.

[52] Cp. Ackert and Church (2001), p. 19.

(1988) study as well as two extensions thereof, both of which are of high relevance for the presentation of this study's design and results in Chaps. 3 and 4.

2.4.3.1 The Smith, Suchanek and Williams (1988) Baseline Market

The Smith et al. (1988) experimental asset market experiment has already been briefly introduced in Sect. 1.4 of the introduction. Due to its central importance for the experiment presented in Chap. 3, it is nonetheless summarized in more depth in the present section.

Smith et al. (1988) conducted experimental market experiments with between nine to twelve traders. The subjects participated in one to three repetitions of a market in which they could exchange assets for cash (and vice versa) in a double auction framework. The maximum length of one period was 240 s, but by pressing a button on their screens subjects could vote to end a period early. In such a case, trading continued either until the last subject had voted to end the period, or until the remaining time in the period had expired without a premature ending. At the end of each period, subjects received a random dividend payout for each unit of the asset they owned. Said dividend was discretely distributed over four equiprobable, non-negative values. The fundamental value process of a unit of the asset has already been graphically illustrated in Fig. 3 in Sect. 1.4 above. Expressed in terms of the number of periods T, the dividend in period t of d_t, and using $E[\cdot]$ as the expected value operator, the fundamental value started out at $T \cdot E[d_t]$ in their experiments and declined by $E[d_t]$ after each period. Since the asset did not bestow any lump-sum terminal payoff, its fundamental value in the last period was just its expected dividend payment for a single period, $E[d_t]$. This fundamental value path was both deterministic and known to all subjects.

One novelty in their design was that all units of the asset (stock) had the same value to every participant, and that all participants could both buy and sell the asset. Prior to their work, experimenters had routinely induced supply and demand schedules characterized by different costs (values) to different designated sellers (buyers) for different units of the asset.[53] The second new design feature in their treatments was that assets did not have single-period lives, but expired only at the end of the experimental session (in their case after 15 or 30 periods).

As noted, Smith et al. (1988)-type experiments almost invariably produce large deviations of transaction prices from the fundamental value, forming bubbles which in some cases even exceed the maximum possible value the asset could ever return in dividends (in the case where only the highest dividend would be drawn in each future period). This is true despite the common knowledge attribute of the

[53] In other words, it cost seller A a different amount to produce a unit of the asset than it cost seller B, and seller A was also subject to different costs for her first and for her second unit. The terminus technicus is that agents faced *heterogeneous reservation prices*. Furthermore, a designated seller (buyer) could not purchase (sell) any unit of the asset in the experiment. See Smith (1976a) for a discussion of induced value theory.

fundamental value process. It seems that some quirk in the process of how subjects form expectations obstructs the market from trading at prices consistent with the underlying fundamental value. As Miller (2002) later put it:[54]

> "At the same time that subjects are learning about the asset's intrinsic value, the market teaches them two things that can undermine that knowledge. First, as the asset price moves toward equilibrium in the early periods, subjects see that prices tend to increase over time. Second, because this increase occurs as the intrinsic value is decreasing, subjects learn that the market price does not need to track the intrinsic value, at least over the short run. Until the markets [sic!] crashes as the experiment nears its conclusion, subjects who learn to ignore the asset's intrinsic value are rewarded by speculative profits, while those who follow it are quickly priced out of the market. Indeed, in experiments that allow selling short, subjects who sell the asset short may not only lose money, should they liquidate their short positions too soon, their purchases can help sustain the bubble."

This and other conjectures regarding the learning process subjects undergo are at the core of the results presented in Chap. 4.

Note that although a number of variations from the original treatment and virtually hundreds of sessions were conducted over the years, the only treatment variation found to reliably and strongly reduce the bubble phenomenon is increased subject experience.[55] The robustness of the phenomenon of market inefficiency in this setting thus provides an exceedingly strong test of the capability of any change in market structure to lead to more efficient information processing. However, once subjects *have* gained experience by participating in repeat rounds, they tend to converge on rational, common, intrinsic dividend value expectations.

As a final observation it should be noted that – despite being often referred to as a stock – the Smith et al. (1988) asset does not bear a high resemblance to the common stocks of most companies. Nonetheless, there are industries where payoffs follow similar patterns as those modeled in the experimental asset markets with declining fundamental value. Good examples could be drawn from investments into the extraction of non-renewable natural resources, such as gold, oil, etc. Depending on the market price for steel for example, an iron ore mine will exhibit random payoffs each period, but will have a fundamental value that declines as the deposit is being used up and approaches zero.

---

[54] Miller (2002), p. 48.

[55] For complete accuracy, this statement needs to be qualified somewhat. First, Noussair and Tucker (2006) demonstrated the complete disappearance of bubbles in an experiment with a complicated structure of futures markets, a setting which unfortunately was rather artificial, thus possessing limited practical relevance (see Sect. 2.4.4.2 of this text). Second, there is mixed evidence with regard to short selling as a means to reduce asset price bubbles, with e.g. Ackert et al. (2001); Haruvy and Noussair (2006) having found evidence for such an effect, King et al. (1993) and Sunder (1995) having reported no such evidence, and Ackert et al. (2006a) having painted a mixed picture (Sect. 2.4.4.5). Third, Davies (2006) found that in a market similar to the Smith et al. (1988) design but with increasing asset values, the experimental asset tended to be undervalued (Sect. 2.4.4.8).

2.4.3.2 The Porter and Smith (1995) Futures Market

Porter and Smith (1995) tested a market design where, in addition to the spot market for an unspecified, dividend-paying good, they enabled the trading of futures contracts on that good in the eighth trading period – the midhorizon point. At the end of period eight, if a trader had a positive net futures position, these accumulated units were transferred to her trading account. If a trader had a negative net futures position, she had to cover the shortfall from her spot inventory.[56] Contrary to previous experiments by the same authors, a period lasted for 300 s, as subjects had to trade in two markets simultaneously. They conjectured that the possibility to trade on the asset's price in the future would facilitate a mechanism of backward induction, leading subjects to refrain from trading the experimental asset at inflated prices.

Porter and Smith (1995) reported that the futures market reduced the bubble amplitude and had no significant effect on duration (defined as the number of consecutive periods in which the difference between mean spot price and fundamental value increased – see Sect. 4.1.2.1) and turnover of the bubble with inexperienced traders in the futures market, but exhibited significantly reduced turnover with experienced futures traders. They interpreted their findings as signifying "that an important function of a futures market is to reduce each individual's uncertainty about other peoples' [sic!] expectations."[57]

Note that the Porter and Smith (1995) futures market was technically a forward market, since the contracts traded in their experiment were not settled daily, but only once at the maturity date.[58]

2.4.3.3 The Lei, Noussair and Plott (2001) No Speculation Treatment

One explanation for the bubble phenomenon in Smith et al. (1988)-type experimental markets is that all subjects are rational, but unsure about the rationality of their fellow subjects. If they assume that at least some other subjects are not rational, even rational subjects might buy the asset at inflated prices in the expectation of being able to resell it again at a later point in time. In so doing, they can earn capital gains and/or dividend income. Conversely, a competing explanation is that subjects simply *are not* rational.

Lei et al. (2001) set out to investigate the distinction between these two propositions. They assigned fixed roles of either buyer or seller to their subjects. A designated buyer could thus only buy assets, but never resell them, while a seller

[56] In the event that a trader had insufficient stock in her spot inventory, she was required to pay a penalty of $4.00, a figure that approximately equals the value of the stock assuming it paid the highest possible dividend ($0.60) in each of the remaining seven periods.

[57] Porter and Smith (1995), p. 525.

[58] Cp. Miller (2002), footnote 11, p. 16.

could never purchase an asset. A rational, risk-neutral or only slightly risk-averse subject being assigned the role of a seller in this design would therefore never sell her assets for a price below their fundamental value. More importantly, no rational buyer would ever purchase shares above their maximum possible dividend value, irrespective of her beliefs with regard to other subjects' rationality. This limits the extent of a bubble to the area between the two stepwise decreasing functions plotted in Fig. 3 (except for the case of clearly risk-loving subjects, a proposition that is itself unrealistic). To investigate also the opposite end of the potential price scale, Lei et al. (2001) employed positive minimum dividend values, such that there was also an absolute lower bound for rational asset sales.

The results Lei et al. (2001) reported were surprisingly clear in rejecting the hypothesis that the observed price patterns could be explained by rational agents who do not possess common knowledge of each other's rationality. They found that between 1 and 16.1% of all transactions (depending on the treatment) in their experiment took place at prices below the minimum, and an impressive 37.6–46.2% took place above the maximum possible dividend value of the asset. The only explanation for this result is the acceptance of the second hypothesis – irrational traders were present in the market. (Table 7 in Sect. 4.1.2.2 reports the relevant analytical bubble measures "Overpriced transactions" and "Underpriced transactions" for Lei et al. (2001) and other studies).

2.4.4 *Alternative Treatment Designs*

2.4.4.1 Dividends and Liquidity

Smith et al. (1988) were the first to depart from the then accepted norm of giving different traders different private dividend values, and found that such different dividend values are not a necessary condition for trade (as many had believed until then). They concluded that there is sufficient intrinsic diversity in subjects' price expectations or risk attitudes (or both) to induce subjective gains from trade. As reported in Sect. 2.4.3.1, their experimental assets paid dividends at the end of each market period, which could take one of four possible and equiprobable values, all of which were non-negative and originated from independent random draws. In some of their experiments they also deviated from their baseline design by paying a final buyout amount for each share to prevent the share price from going to the expected value of a single dividend draw in the last period. This buyout value equaled the sum of the dividend draws over all 15 periods plus or minus a constant (each with probability 0.5). The aim of this institutional detail was to enhance the possibility of a bubble; a measure that proved unnecessary and was subsequently dropped.

Smith et al. (2000) examined markets modeled after the example of the Smith et al. (1988) markets, but with three different dividend treatments. In the first institution, the asset paid a single dividend at the end of the trading horizon. In the second, dividends were paid at the end of each trading period (as in the classic

Smith et al. (1988) design). In the third setup, the asset paid some dividends at the end of the trading horizon and some at the end of intermediate periods. They found that the second setup produced the strongest bubble phenomenon. While the third design also reliably generated bubbles, they were less pronounced. The first treatment, finally, yielded a bubble in only one of ten sessions, suggesting that frequent dividend payments were conducive to the formation of bubbles. This result underlines the important role dividends play for the formation of bubbles in experimental asset markets – a role that will also be discussed before the background of the results in Chap. 4.

Noussair et al. (2001) conducted another experiment probing the role of dividend payment frequency and fundamental value structure for the observed market prices. They employed a dividend pattern of four discrete dividends with an expected value of zero, complemented by a lump-sum terminal payoff.[59] In contrast to the declining values of assets in earlier experiments, in this new structure the expected value of one unit of the asset remained constant and equal to the value of the terminal payoff throughout the experiment. With this setup, they found that bubbles occurred in only four out of eight sessions and exhibited smaller magnitudes, a marked improvement in market efficiency over the baseline markets. Nonetheless, the fact that the market exhibited a bubble pattern in 50% of all rounds proves that the frequently changing (and monotonically declining) fundamental value of earlier designs is not a necessary condition for bubble formation. In a related experiment, Davies (2006) modified the dividend structure so that the expected dividend became negative, and introduced a terminal payoff for each share of stock. This led to an asset exhibiting increasing fundamental value, which he found to cause trading at prices considerably below the fundamental value. He conjectured that the reason for this inversion of the observed price deviation may have been due to both failure of the agents to upwardly revise their perceptions of value over time and to decreasing liquidity relative to fundamental value as the experimental round progressed.

Porter and Smith (1995) investigated the role of risk in experimental markets. In their setting, the traded good paid future dividends which were certain, thereby eliminating from the experiment both risk and the influence of varying degrees of risk aversion among subjects. They found that the elimination of dividend risk had no significant effect with inexperienced traders. They also tried to confront subjects twice experienced in the certain dividend environment with a risky dividend structure, but failed to rekindle a price bubble. More specifically, they reported that their results were indistinguishable from experiments with traders twice experienced in a risky dividend environment. Finally, the certain dividend structure did not significantly reduce the bubbles observable in similar markets with risky dividends.

[59]The four equally likely dividend values in their experiment were -24, -16, 4, and 36 units of experimental currency, while the terminal payoff consisted of 360 units. They referred to the two low values as *holding costs* to explain their negativity.

Van Boening et al. (1993) wanted to focus subjects' attention on the asset's expected value in an experimental double auction market, hoping that this would lead to less prominent deviations of market prices from fundamental values. To test this hypothesis, they departed from the common design of a discrete dividend distribution with four asymmetric and equally likely points (e.g., 0, 4, 8 or 20 cents, all with equal probability of 0.25, as in design 1 of Smith et al. (1988)) and used a discrete distribution with five symmetric points with unequal probabilities (5, 15, 25, 35, and 45 cents with probabilities 1/9, 2/9, 3/9, 2/9, and 1/9, respectively). Unfortunately, their results did not show a decline in the propensity of experimental markets to produce asset price bubbles under this new dividend regime. Caginalp et al. (1998) similarly modified the dividend structure in their experiment, where subjects traded a stock with a single dividend payment, payable in the last of 15 periods. The participants knew that the dividend had a 25% chance of being either $2.60 or $4.60 and a probability of 50% of being $3.60, implying an expected value of $3.60. Unfortunately, Caginalp et al. (1998) focused on the role different ratios of cash versus share value played in the price process, but did not report on how their experiments compared to standard Smith et al. (1988)-type markets.

Oechssler et al. (2007) ran experimental markets where the subjects could trade five different assets simultaneously, all of which paid a single dividend at the end of the experimental session. In order to find whether subjects were aware of overpricing but speculated on even higher prices, or whether they were unaware of deviations from fundamental values, the experimenters asked them to predict both the period end price and the final dividend of an asset. Their findings were consistent with subjects who were aware of overpricing, since they provided final dividend estimates in line with fundamental values. At the same time, they forecast prices significantly exceeding fundamental values in periods with bubbles. The authors also found that bubbles can occur without intermittent dividend payments if – as was the case in their experiment – inside information is present in the market, while finding no evidence for bubbles in a similar setting without insiders or with traders who were provided with the means to communicate. In fact, they found that the number of messages sent in sessions with a chat function was negatively related to the frequency of bubbles. They conjectured that an explanation for this could be that communication provided the means for more sophisticated traders to "educate" their less sophisticated colleagues, and to accelerate the synchronization of expectations. Furthermore, they also pointed out that the possibility to communicate might have given subjects something other to do than trading – a hypothesis that is consistent with the Active Participation Hypothesis suggested by Lei et al. (2001). The Active Participation Hypothesis implies that irrational actions in laboratory experiments may be due to the fact that subjects are required to participate in an experiment until the end and have no other activity available to them than that of acting in the experimental market. Colloquially speaking, subjects may "act out of boredom," instead of out of a desire to improve their payoff from the experiment.

Ackert et al. (2006c) employed two different assets, each of which, at the end of a period, offered a zero payoff with a probability of 0.98 and a payoff of $20 with a probability of 0.02. The only difference between them was that – within one experiment – the first could pay out the dividend of $20 an unlimited number of times, while the second only paid out $20 if there had been fewer than three dividend payouts of $20 in earlier draws. The authors referred to the second kind of asset as "truncated." Due to the low probability of a payout, the expected values of these two assets were virtually identical. Aiming to prove that traders are subject to probability judgment error and irrationality, Ackert et al. (2006c) used these assets in three types of markets: One with ten periods and dividend draws after each period and a second (third) with a single period and eight (five) dividend draws, all of them after the single market period. They found that the difference in prices of the untruncated asset and the truncated asset was generally positive and declined as the experiment progressed. They also reported a positive correlation between the magnitude of the difference in prices of the untruncated asset and the truncated asset and the occurrence of bubbles. Furthermore, median asset prices in multi-period markets were higher than corresponding prices in single period markets in all cases. Similarly, the difference between prices in multi-period markets and single period markets was larger when there were eight dividend draws than when there were only five, which indicates that subjects engaged in speculation. Finally, they showed that the magnitude of the difference between asset prices in multi-period markets and single period markets was considerably greater in bubble markets than it was in non-bubble markets, another indication for speculation activity.

In a twist on the experiments just described, Ackert et al. (2006a) investigated the effects of margin buying and short selling on experimental asset markets with two assets – one with standard and one with lottery characteristics. They found that in markets with margin buying but without short selling, bubbles could be observed for both assets, with the lottery asset exhibiting the larger bubble. Restricting margin buying dampened the bubbles and caused the difference in bubble size between the two assets to disappear. When they restricted margin buying and allowed short selling, they did not observe bubble-and-crash patterns. Finally, consistent with Haruvy and Noussair (2006), they found that in some markets the lottery asset traded at prices considerably below its fundamental value.

Ackert et al.'s (2006a) results hinted at a connection between subjects' ability to buy on margin and the bubble extent. More generally, bubble extent seems to increase with increasing liquidity (i.e., the ratio of cash to stock value) in experimental asset markets. An article that provides evidence of this nature is Caginalp et al. (2001). They reported that each dollar per share of additional cash in their experimental markets (with periodic dividend payments) was associated with a $ 1 increase in the maximum share price, a $ 0.45 increase in the average transaction price and a $ 1.11 increase in the maximum price deviation from fundamental value. These effects were considerably reduced when subjects received information on all outstanding bid and ask quotes, i.e., when there was an open order book (the corresponding figures were: $ 0.36 increase in maximum share price, $ 0.28 higher average transaction price and $ 0.32 larger maximum deviation).[60]

[60] Cp. Caginalp et al. (2001), p. 87.

Note that the connection between the treatments with margin buying and those with dividend payments lies in the fact that both increase the amount of cash subjects can spend on stock purchases – an observation that is made in Huber et al. (2008).

Summarizing the findings of the studies discussed above, the verdict on the role of liquidity, and specifically dividends, in experimental asset markets is one of far-reaching importance. Apart from subject experience, dividends are the parameter with the most visible impact on the formation of bubbles. Due to their importance, they are also accorded some space in the discussion of the results in Chap. 4.

2.4.4.2 Futures Market

Early in the 1980s, experimental economists started investigating the influence of futures markets on spot market prices. These first experiments separated spot and futures trading periods and varied the asset's fundamental value. After the publication of Smith et al. (1988), experimentation turned to operating spot and futures market simultaneously, a design that more closely resembles real-world markets. Since this branch of experimental research constitutes the first experimental evidence on the effect of derivative markets on spot markets – the first option market experiments were conducted later – this literature is briefly reviewed in the following paragraphs.

Forsythe et al. (1982) conducted five oral double auction asset market experiments, where trading was structured into six to eight years, with two periods (A and B) each. All period A's of a given market were identical with respect to the underlying distribution of asset returns and all period B's were also identical (although different from the period A's of each year). In one of their five markets, a futures market of period B assets operated in period A and replaced the spot market of period B's assets. The results led the authors to conjecture that the existence of a futures market may increase the speed of information dispersal as well as the convergence to equilibrium, might remove the necessity of replication, and could increase market efficiency. Due to the small sample size of a single experiment, however, their results can only be interpreted as weak support of these conjectures. The same authors reported on nine additional experimental asset markets in Forsythe et al. (1984), four of which followed the original spot-only structure, with the remaining five featuring spot-and-futures trading. Their new results were less ambiguous and strongly confirmed the conjectures from the paragraph above. While the hypothesis that prices were transacted in the range predicted by a rational expectations equilibrium could not be rejected at the 5% significance level in 17 out of 35 years for the markets with futures trading, in the spot-only markets it was rejected 27 out of 28 times. They found that futures markets did accelerate convergence and that in the absence of futures markets, even experienced traders had problems overcoming the existing coordination problems. Interestingly, Forsythe et al. (1984) noticed that spot prices exhibited considerably increased variability in the early market years if there was a futures market present.

They explained this finding by stating that, since futures prices play a role in publicizing private information, the higher variance was a sign of the higher speed of convergence toward the rational expectations equilibrium. Testing this intuition analytically, they found that in the first 3 years, where the group of experiments with a futures market converged rapidly to rational expectations equilibrium prices, the group of experiments with futures markets always had significantly more price variation than those with sequential markets. Performing pair-wise comparisons between experiments with futures markets and those with sequential markets, but with otherwise identical parameters, they found that the hypothesis that the variance in the first case is larger than that in the second could be rejected at the 5% significance level in only six out of 70 cases.

Friedman et al. (1983) conducted four experimental asset markets, each with three trading periods per market year, referred to as periods A, B and C, with identical certificate returns across market years. Two of the four markets permitted only spot trading; the other two markets featured trading of spot contracts and futures contracts for period C-delivery in the two periods A and B, with no trading (but delivery of the futures contracts' underlying certificates) in period C. In all experiments, traders received a trading commission of one cent per transaction. Their results showed that the first (second) periods of experiments with futures trading converged slightly (considerably) faster than the spot-only markets. The evidence also suggested that the standard deviation of transacted prices was smaller in markets with futures trading than in those without. The authors interpreted these findings as supporting the conclusion that futures markets were associated with informationally more efficient spot market prices. In Friedman et al. (1984), the same authors reported that the hypothesis of a lower coefficient of spot price variation for a spot-and-futures treatment than for a spot-only treatment received a significance level of only 34.7 (29.8) percent in period A (B) in an environment of certainty. In the case of uncertainty regarding the future state of nature, the results were considerably stronger, with a significance level of 5.4%. Pooling the final year data across all their six experiments (and all periods) run with experienced subjects, they obtained a 19.7% significance level for a reduction of the coefficient of variation in the presence of a futures market. They criticized the results of Forsythe et al. (1984) on the grounds that the latter employed a joint treatment variable mixing the effects of the addition of a futures market to the spot market with those of trader experience. This led them to conclude that the results of Forsythe et al. (1984), which had suggested a higher coefficient of spot price variation in the presence of a futures market, also supported the results of Friedman et al. (1984), reporting a lower coefficient of variation. Friedman et al. (1984) also reported that insiders in their markets earned higher profits than non-insiders in every market year, yet this effect was reduced (with a significance level of 27.4% in a Mann-Whitney test) by the presence of a futures market, which they interpreted as evidence that the futures market caused "leakage" of insider information. Finally, they reported that futures markets tended to speed up the evolution of prices to more informationally efficient equilibria in the case of uncertainty regarding the future state of nature.

These experiments, which were conducted prior to the work of Smith et al. (1988), formed a blueprint for applying a similar analysis to their new design. In other words, they suggested the question of how adding a futures market would change the results in a Smith et al. (1988)-type environment. The first to address this question were Porter and Smith (1995). As reported in Sect. 2.4.3.2, they found that a futures market reduced the amplitude of the price bubble in the spot market and had a similar effect on the turnover measure for rounds played with experienced subjects, yet did not succeed in eliminating bubbles altogether. This accolade went to a treatment ran at the University of Canterbury, NZ, and at Purdue University, USA, in late 2002 and early 2003. Using an ingenious experimental market structure, Noussair and Tucker (2006) forced their subjects to form expectations about future prices by backward induction. In addition to a normal 15-period stock market, they operated 15 futures markets, each maturing at the end of one of the 15 trading periods. To prevent their subjects from being distracted from the backward induction task, they first opened only the period 15 futures market for trading. After a fixed pre-announced time interval, the period 14 futures market opened, and so on. Only when all 15 futures markets were open did the spot market start operating. At the end of the first period of spot trading, the period 1 futures market (i.e., the last futures market that had been opened) matured and was closed. They found that futures market prices deviated considerably from those suggested by rational expectations, but converged to levels close to the latter as they approached their respective maturity dates. More importantly, spot market prices closely tracked the stock's fundamental value. Unfortunately, this remarkable success in completely eliminating the bubble phenomenon came at the expense of an inherently artificial market structure, which does not lend itself to application in real-world markets.

2.4.4.3 Option Market

Experimental research findings from studies on the impact of option markets on spot market prices are the closest analog available for comparison to the work presented in this text. The largest discrepancy between earlier designs and this study's treatments is that prior work used conventional (usually European) options, while this study employs the digital option contract used in some online prediction markets. The literature on asset market experiments offering the possibility of option trading is briefly reviewed below.

Biais and Hillion (1994) analyzed the impact of the introduction of a non-redundant option into a double auction market populated with noise traders and an information trader (in their model called liquidity traders and insider, respectively). They found that option trading sometimes reduced the profits of the insider, yet did not do so reliably (i.e., for all parameterizations). Furthermore, they wrote that the introduction of the option seemed to mitigate the problem of market breakdown. Such a breakdown occurred when noise traders perceived the transaction costs due to asymmetric information (i.e., the risk of being exploited by better-informed insiders) to outweigh the possible benefits they could attain from trading to improve their asset-to-cash ratio. They conjectured that – by making the market more complete – the option reduced the risk from asymmetric information and thereby also reduced the frequency of market breakdowns.

Kluger and Wyatt (1995) conducted oral double auction asset market experiments and designed treatments somewhat similar to those used here. They ran one treatment only with an asset market and one where the asset market was complemented by an option market operating sequentially, with trading alternating between the two markets. Their findings in this setting showed that options dramatically accelerated the information aggregation process, making the asset market informationally more efficient. They also stated their impression that the efficiency gain was due to the options enriching the message space and speeding up the discovery of the correspondence between signals (about fundamental value) and prices.

De Jong et al. (2006) ran an experimental asset market and an option market, wishing to determine whether the presence of an option market would improve the market quality of the underlying asset by leading to price discovery across both markets. In their experiment, three competing dealers in each market were the counterparties to both the single existing insider – who knew the intrinsic value of the asset – and to two liquidity traders. The authors did not impose borrowing or short-sales constraints, so that leverage effects, which might have made options attractive to informed traders in real markets, were absent in their experimental treatments. All trades were constrained to a lot size of a single unit. The liquidity traders were faced with exogenous liquidity shocks by being required to meet uncorrelated end-of-period positions in both the option and the underlying.[61] They found that price efficiency in the asset market was higher and the asset's price volatility lower when the intrinsic value of the option was positive and that the presence of an option generally improved market efficiency (i.e., even if its intrinsic value was zero). They also reported that the insider, who could choose between trading in the market of the underling or in that of the option, chose the more profitable market to trade in 86.3% of all cases. Price discovery thus took place in both markets, and market markers in the asset (option) market revised their quotes in the direction suggested by the situation in the option (asset) market.

2.4.4.4 Monetary Incentives

Utility theory does not predict that people will make the "correct" decision when it is not in their interest to do so.

Vernon L. Smith (1973)

Over the set of all economics experiments in the literature, a large number of compensation mechanisms has been employed. Some experiments used hypothetical

[61] In addition to posting bid and ask quotes, all market makers could also initiate transactions with other market makers in either market. They neither received information regarding the end-of-period value of the asset, nor were they told the required end-of-period positions of the liquidity traders.

payoffs, some real payoffs, some converted the currency used in the experiment into real money, while others used real money in the experiments. There were experimenters who paid the full earnings for all decisions to their subjects, while others chose individual decisions or rounds at random and rewarded only those. This section gives an overview of the differing monetary incentive schemes employed in the literature. In the process, it also attempts to summarize findings on the impact these different incentive schemes had on experimental results.

Smith (1962) first made his observations regarding the behavior of double auction markets in a setting using hypothetical payoffs, and only later confirmed his results in real money markets. In Smith (1965), he again took up this line of research and investigated the differential impact of full payoffs (every subject received the payoff she had earned over the course of the experimental session) versus random payoffs (only a randomly chosen subset of all subjects received the payoff they had earned). He reported that in the random payoff treatment, actual equilibria deviated significantly more strongly from the theoretical equilibria than in the treatment of full payoffs.

Smith and Walker (1993a) investigated the bidding behavior in first price auctions with varying payoff levels. They reported that increases in real payoffs led to higher bids and fewer decision errors. Following up on their first article, Smith and Walker (1993b) conducted a survey on experimental articles that reported on the comparative effects of subject monetary rewards. They found that the error variance of observations around the predicted values tended to decline with increasing monetary rewards. From their observations they derived a theory of decision making that is a function of the effort dedicated to the decision-making process, an approach that is in line with earlier observations by Smith regarding the subjective cost of transacting.[62] The higher cognitive and response effort that is necessary for decisions closer to the optimum entails a disutility that can be compensated by higher monetary incentives. Nonetheless, for a given decision problem, higher monetary rewards might remain ineffective if – due to the complexity of the decision task – the agent's maximum possible effort has already been reached. The authors refer to this conjecture as a "labor theory of decision making."

A comprehensive survey of articles on the effect of financial incentives in economic experiments is Camerer and Hogarth (1999). In their deliberations, they distinguished "declarative knowledge," i.e., knowledge about facts, from "procedural knowledge," i.e., skills and strategies for using declarative knowledge in problem solving. Based on a literature overview of 74 experimental studies they concluded that subjects learnt from observation and "by doing," as opposed to "by thinking." However, they observed that effortful thinking can substitute for a lack of cognitive capital (i.e., declarative knowledge) in some tasks. As an example they quoted the stagecoach problem, which involves finding the least-cost series of nodes connecting two nodes in a network. Subjects with cognitive capital in the form of knowledge about the dynamic programming

[62] Cp. Smith (1985), p. 268, and as quoted therein, Smith (1982), p. 934.

principle were found to be able to backward induct and solve the problem with little effort. Subjects without such knowledge solved the problem with much larger effort by brute-force trial-and-error.

Relating this to monetary incentives, they found that incentives were ineffective in situations where the marginal return to increased effort was low, which was the case whenever it was either very easy or very hard to do well. In the first case, the monetary incentives did not matter because they were unnecessary to induce good performance; subjects did well even without incentives (or because they had sufficient intrinsic motivation). In the second scenario, even though subjects were incentivized to do their best, their effort failed to achieve significant improvements in their performance because the task was too hard or too complex for the experimental agents. However, in intermediate situations, the argument from the last paragraph seemed to become relevant – where increased cognitive effort was able to improve performance, monetary incentives sometimes caused better outcomes. This was also recognized by Hertwig and Ortman (2001), who wrote that:[63]

> "[...] economists think of 'cognitive effort' as a scarce resource that people have to allocate strategically. If participants are not paid contingent on their performance, economists argue, then they will not invest cognitive effort to avoid making judgment errors, whereas if payoffs are provided that satisfy saliency and dominance requirements [...], then 'subject decisions will move closer to the theorist's optimum and result in a reduction in the variance of decision error' [...]"

As a case in point, the authors quoted earlier studies which had found positive incentive effects in settings requiring little skill, such as pain endurance, vigilance or clerical or production tasks, while reporting weaker effects in memory, judgment, and choice tasks and no positive (and sometimes negative) effects in experiments involving problem solving. Nonetheless, even in situations where incentives had failed to improve performance, they had frequently decreased the variance in subjects' performance. If aggregate behavior is sensitive to outliers, which in turn are sensitive to monetary incentives, this induces a causal link between incentivization and aggregate results.

Moving away from the theory of cognitive effort, Forsythe et al. (1982) were among the first to use an artificial currency in experimental markets. They argued that using dollars (i.e., real currency) would have been prohibitively expensive in their experiments, where they distributed initial cash positions of between 10,000 and 20,000 currency units. They converted their artificial currency "francs" into dollars by calculating the payoffs for a given year as $a + bx$, where x was the quantity of francs held by a subject at the end of a trading year, $b > 0$ was a factor for the conversion of francs to dollars[64] and $a < 0$ were fixed costs which approximately equaled the initial cash endowment. In response to this, Friedman et al. (1983) somewhat disparagingly referred to the Forsythe et al. (1982) franc as an "arbitrary

[63] Hertwig and Ortman (2001), pp. 25–26.

[64] Conceptually, b is the exchange rate of dollars for francs, which in their experiments was set to \$0.002 per franc.

unit of account," writing that in their own article they "avoided what seems to us the needless complication (for traders) of converting francs to dollars."[65]

Ang et al. (1992) offered significant additional bonus payments to those of their subjects earning the highest profits in the first periods of a two-period-asset experiment, with the aim to shorten their investment horizons along the lines of portfolio managers in the investment community. They found that this modification caused a large bubble in the first periods of their market, with prices in the second periods remaining close to the risk neutral equilibrium. Since this bubble was only partially reduced when traders had to invest $20 of their own (real) money, they conjectured that its reason lay in the imbalance between the buying and selling powers of subjects in experimental markets, which in real markets are reflected in short-sale restrictions and high costs, as well as in the possibility to leverage long positions. Modifying this design by increasing the asset endowment and decreasing the cash endowment to approximately the market value of assets led to a disappearance of the bubbles and caused trading to take place at a discount from the risk neutral equilibrium in the first periods.

In another experiment employing a compensation scheme non-linear in terminal wealth, James and Isaac (2000) tested whether tournament incentives (i.e., compensation that is strongly dependent on an individual's outperformance of the average market participant, a common attribute of mutual fund managers' payoff functions) changes the common bubble-and-crash pattern in markets following the structure of Smith et al. (1988). They found that even for subjects who had previously participated in (at least) two markets without tournament contracts (and who were therefore expected not to produce any more bubbles), the repeated imposition of tournament contracts led to increasing deviations from fundamental value pricing, thus underlining the impact even comparatively small changes in the compensation scheme can have on experimental results. Williams (2008) also employed a rank-order tournament incentive scheme, awarding extra credits to student subjects who ranked best in final experimental cash holdings in their asset markets, but did not report on the effect of this institutional detail on the experimental outcomes.

Luckner and Weinhardt (2007) ran a prediction market for the FIFA World Cup 2006 with three different payment schemes to test a similar proposition as the two articles discussed in the previous paragraph. A first group of 20 students was paid a fixed amount, the three best-performing traders of a second group of 20 subjects received a payoff related to their rank within the group (with the remaining 17 players receiving nothing), and a final group of 20 players received a payoff that depended linearly on their terminal wealth. The average payment per agent was held constant (at € 50) over all three groups. They found that the third group (which was being rewarded according to what they termed a "performance compatible payment" scheme) actually yielded market prices that corresponded to predictions which were worse than randomly drawing one of the three events the prediction

[65] Friedman et al. (1983), p. 130.

market was meant to forecast. Conversely, the rank-order treatment outperformed the other two payment schemes and even the fixed payment group did better than the third group of subjects. The authors conjectured that their subjects were motivated by factors extrinsic to the experiment as opposed to the monetary incentives, but did not conduct any control experiments to alleviate the problem of their small sample size.

Ackert et al. (2006b) did not focus on the level of actual or expected payoffs, but instead analyzed the path dependence of subject actions conditional on the development of their wealth. In their investigation of the house money effect, they found that not only the expected payoff, but also payoffs received earlier in the experiment influence behavior in asset markets – the phenomenon that individuals tend to become less risk-averse after having recently received a gain. Their results from nine experimental sessions, with eight subjects each, showed that they could indeed observe the house money effect in their laboratory experiment. It is specifically this last experiment that shows that the word is not yet in on how compensation affects behavior in and results of asset market experiments, and that more sophisticated models and tests are needed to analyze this topic.

2.4.4.5 Short Selling

Experimental asset market bubbles are caused by subjects willing to pay exaggerated prices for the experimental good. Several studies reported that seemingly irrational traders kept trading at exaggerated prices long after more rational subjects had run out of assets. Due to this lack of liquidity, the latter were rendered unable to contribute to bringing prices back into line with fundamental values. Theorists thus conjectured that if subjects were permitted to sell short, more rational traders could profit from the asset's overvaluation by selling it. Such sales would at the same time keep prices at lower levels, because the irrational subjects would not have to trade only among themselves, but could instead enter into transactions with (rational) traders offering units of the asset for a lower price. This is a simple supply-and-demand equilibrium argument where increasing supply leads to lower prices. In an early experimental study contradicting this conjecture, Sunder (1995) reported that short selling in their experiment did not reduce the number of periods experimental asset market bubbles lasted, nor did it decrease their size. While this result was not encouraging for proponents of the efficient market theory, a much more alarming result was published in Haruvy and Noussair (2006). Ernan Haruvy and Charles Noussair employed a market similar to that of Smith et al. (1988) to study the effect of the relaxation of short-sales constraints on the bubble phenomenon typical for this market structure. They motivated their research with the observation that "In the absence of short selling, the asset price will simply be the price offered by the most optimistic trader with sufficient funds."[66] Contrary to King et al. (1993) and in line with Ackert et al. (2001) they found that decreasing obstacles to short-selling was

[66]Haruvy and Noussair (1993), p. 1155.

associated with lower security prices. Yet while Ackert et al. concluded that increased short-selling capacity led to more efficient markets, Haruvy and Noussair increased the extent of possible short sales in further experiments and found that facilitating short-selling seemed to simply decrease prices. When they permitted traders large leeway in their short transactions, prices followed a negative bubble pattern, consistently remaining below the fundamental value over the course of the experiment. They conjectured that the increased availability of units of the asset (i.e., a higher supply) paired with constant demand led to a decrease in the observed price.

Haruvy and Noussair (2006) also confirmed for their own setting a result from Caginalp et al. (2000b), who had reported that increasing the cash available for asset purchases increased transaction prices in settings without short-selling. Despite their modifications aimed at making the market more efficient, the markets in the Haruvy and Noussair (2006) experiments exhibited very high transaction volumes, large price swings relative to fundamental values and long periods of trading away from the asset's fundamental value.

2.4.4.6 Variations in the Subjects Variable

An important question in the experimental economic science is whether experiments with student subjects yield the same results as ones using business professionals. To investigate this issue, Dyer et al. (1989) compared the performance of upper-level students majoring in economics ("naive agents") with that of experienced business executives from the construction contract industry ("experts") in a laboratory experiment where participants were bidding for contracts, subject to an uncertain cost structure. They found that both subject populations exhibited irrational behavior and were subject to the winner's curse, with no significant differences at the 10% level or better in any of the following performance measures: the proportion of times the low bid was submitted by the agent with the lowest cost signal, average actual profits, the proportion of times the low bid was less than the rational minimum amount, and the proportion of times the low bid was less than the rational minimum amount at the individual level. (Conversely, Alevy et al. (2007) documented that in their information cascade experiments, market professionals emphasized their private information more strongly than did student subjects and were also impervious to which domain of earnings – gains or losses – they were operating in.) While there was no evidence for behavioral differences attributable to the subject pool, Dyer et al. (1989) did find some differences in behavior that they attributed to heterogeneity in risk aversion. They wrote: "The different pattern of profits/losses [. . .] and the differences in estimated bid functions, lead us to reject the maintained hypothesis that there are *no* differences between the two subject pools; however, we feel that the similarities are much more striking than the differences."[67] Güth et al. (1997) also specifically analyzed the impact of subjects'

[67] Dyer et al. (1989), p. 112.

risk aversion. They found that risk aversion – as determined in a pre-test to their experiment – had no explanatory power for the subsequent portfolio choice in a multi-period capital market experiment. In a purely descriptive article regarding the risk appetite of different types of subjects, Faff et al. (2008) surveyed a number of studies on this topic. They found that risk tolerance increased with education, income and wealth, decreased with age and was lower for females than for males and for married than for unmarried investors.

Ackert and Church (2001) compared results from experiments run using only senior business students as subjects with experiments conducted with only freshman arts and sciences students who had their majors outside of the fields of business and economics, and with a third type of subject pool formed from mixtures of the two groups. They found that bubbles were reduced when business students gained experience, while the same was not true for the non-business students.[68] Furthermore, experienced business subjects were able to make profits at the expense of inexperienced subjects from both subject pools. They also let their subjects forecast prices at the beginning of each period and found that in markets with business students, superior forecasters outperformed other traders in terms of profits. Ackert and Church (2001) summarized their results by stressing the importance of considering agent type in the development of models characterizing economic behavior.

In his experiments conducted at Indiana University, Williams (2008) modified not the subject pool but the size of the sample he drew from it. He reported on three asset market experiments, run over 8 weeks, and using between 244 and 310 traders. In these experiments, all agents were endowed with the same number of shares of stock and with the same amount of experimental currency, and they could access the market software at any time over fifteen periods, the majority of which lasted for 3.5 days. At the end of each round, owners of a share of stock received a common dividend stemming from a rectangular distribution. Extra credits were then awarded to the best subjects using a rank-order tournament design. Students participating in these markets were encouraged to discuss it with one another. In addition, the interim results of the markets were discussed in class during their operation. Surprisingly, the experiment yielded results very similar to those of comparable markets conducted with much fewer traders, a monetary reward structure, and in the laboratory.

2.4.4.7 Institutions of Exchange

The striking competitive tendency of the double auction institution, which has been confirmed by at least a thousand market sessions in a variety of designs, indicates that neither complete information nor large numbers of traders is a necessary condition for convergence to competitive equilibrium outcomes.

Charles A. Holt (1995)

[68] Ackert and Church (2001), p. 18.

The basic institution of exchange in many markets in all kinds of settings is the auction, a transaction medium that has been employed by mankind for millenia. Herodotus, in the fifth century B.C., described how women were auctioned off to be wives in Babylonia; in the Roman Empire, booty was transferred via auctions; and the possessions of deceased Buddhist monks in seventh century China were allocated to new owners using the auction institution.[69] Naturally, the technology of auctions has evolved since their first application in early human history, and today encompasses a variety of forms. Since the specific form of the auction mechanism is an important determinant of trader behavior and allocational efficiency in a market, some evidence on different transaction mechanisms is presented in the following paragraphs.

Smith (1976b) presented five institutions of exchange: the double auction, the bid auction, the offer auction, posted pricing, and the discriminative and competitive sealed-bid auctions. In a double auction, buyers and sellers submit bids and asks, which are tabulated and compared. When a buyer (seller) submits a bid (ask) which equals or exceeds (equals or is smaller than) the lowest ask (highest bid) in the market, a transaction takes place. In a continuous double auction, an order book is maintained and auctioning continues after transactions take place. The bid (offer) auction is similar to the double auction, with the difference that only the buyers (sellers) may post price quotes, while sellers' (buyers') single possible action is to accept a bid (offer). In a posted pricing market, sellers (buyers) independently select reservation price levels, which are then communicated to the market. Next, a buyer (seller) is chosen at random and matched with a seller (buyer), whom she can then make an offer at that seller's (buyer's) posted price. This procedure is repeated until the initial buyer (seller) does not demand any additional units, at which point a new buyer (seller) is chosen at random. Finally, in the discriminative (competitive) sealed-bid auction, the seller offers a specified quantity of the good and buyers submit bids. These are sorted highest to lowest and the highest bids are accepted, such that the seller's quantity can be fully allocated. The transaction price is the full price bid by the buyers (the price of the lowest accepted bid) in the case of the discriminative (competitive) sealed-bid auction.

Smith (1976b) reported that, in the double auction, "prices converge to 'near' the theoretical (Supply=Demand) equilibrium level usually within the first twenty to thirty transactions."[70] He furthermore wrote that the quantities exchanged were usually within one unit of the theoretical equilibrium, that an order improvement rule – requiring that new bids (offers) improve on the currently outstanding best bid (offer) – did not significantly accelerate convergence, and that convergence tended to be from below (above) when the producer surplus was larger (smaller) than the consumer surplus. Considering the variations of the bid and offer auctions, Smith (1976b) found that the side having the pricing initiative was usually disadvantaged with regard to eventual transaction prices, while in a posted-bid environment, the

[69] Cp. Milgrom and Weber (1982), p. 1089.

[70] Smith (1976b), p. 48.

opposite tended to be the case. Finally, accepted bids in competitive sealed-bid auctions stochastically dominated (i.e., were higher than) bids in discriminative sealed-bid auctions. Smith et al. (1982) built on this earlier work and also compared five market exchange institutions: the double auction (DA), a sealed bid-offer auction mechanism (PQ), a variable quantity sealed bid-offer auction mechanism (P(Q)), and tâtonnement versions of PQ and P(Q), referred to as PQυ and P(Q)υ. In the PQ mechanism, buyers (sellers) submitted a maximum bid (minimum ask) price and quantity, and an algorithm then determined a single market clearing price. In the P(Q) treatment, each buyer (seller) submitted one bid (ask) price for each unit of the asset she was assigned a valuation for, and the same algorithm as in PQ was used to determine a single market-clearing price. In the tâtonnement treatments PQυ and P(Q)υ, each trader had to give her consent to a proposed price and allocation offer. If there was a consent before the maximum number of trials T was reached, T times the proposed bid and offer quantities were exchanged. If there was no consent, no trade took place. Smith et al. (1982) found that the DA treatment yielded higher overall efficiencies than the PQ mechanism, even though experience seemed to ameliorate this difference. PQ in turn did not turn out to be inferior to PQυ, which yielded prices that were as erratic as under the non-tâtonnement institution. The P(Q) mechanism outperformed PQ, but underperformed DA. However, its tâtonnement version, P(Q)υ, performed at least as good as the double auction. Similarly, Pouget (2007) compared the performance of a call market and a Walrasian tâtonnement, making sure that both market institutions had similar equilibrium outcomes in both prices and allocations. He found that the gains from trade were higher in the Walrasian tâtonnement institution than in the call market, despite the fact that prices were fully revealing in both markets. Uninformed traders did not participate in the call market to the extent predicted by theory, a fact that Pouget (2007) traced to bounded rationality and strategic uncertainty. He wrote:[71]

> "Overall, this paper shows that limitations on human cognition can create transaction costs. Yet, adequate design of the market structure can overcome the impact of cognitive limits. In this experiment, compared to a Call Market, a Walrasian Tatonnement provides a way to economize on cognitive transaction costs. I explain the greater performance of the WT in terms of more tractable mental representations and robustness to strategic uncertainty, both features which foster learning. Hence, this paper suggests that even when it does not influence strategic outcomes, market design may still be an important source of efficiency gains through its effect on traders' ability to discover equilibrium."

Cason and Friedman (1996) conducted 14 laboratory experiments on double auctions, finding them to be a very efficient market structure and noting that initial inefficiencies (i.e., arbitrage opportunities) disappeared with increasing experience of market participants. Van Boening et al. (1993) compared a conventional double auction market setting with one that used call auctions, and expected a reduction of the bubble phenomenon that is well-documented for the former setting. Their results did not confirm their expectations, but showed that the change in trading

[71] Pouget (2007), pp. 303–304.

institution did not eliminate the bubble phenomenon. Haruvy et al. (2007) also documented bubbles in a call market setting. Liu (1992) found that in her experiments, continuous double auctions outperformed call auctions in terms of efficiency when all traders were endowed with diverse information, while the opposite was true when uninformed traders traded alongside diversely informed traders.

Easley and Ledyard (1993) were the first to work on a positive theory of how prices are formed and of the trading process in an oral double auction market. They derived their model from some ad hoc assumptions not flowing from an optimizing model but rather based on observations of empirical behavior of market participants, and reasonable interpretation of their actions. In a next step, they applied their theory to a number of empirical experiments both from oral and from computerized laboratory double auctions. Their predictions were largely borne out by the evidence, even though there were a small number of deviations in every experiment. In an even more universally applicable account, Jackson and Swinkels (2005) provided a general proof of the existence of at least one equilibrium involving positive volume of trade for double private value auctions.

Crowley and Sade (2004) investigated what effect the option to cancel orders has on trading volume and prices in a double auction environment. In their design, subjects could post one bid and one ask at a time in a continuous double auction market operating over 12 periods, lasting 3 min each. They conducted experiments using two different treatments – one in which traders could cancel their bids and asks, and one in which they could not. In the former, they found that the mean portion of orders that were being canceled was 4.2%, and that the mean number of standing orders was 46.52 versus 29.3 in the treatment without cancelation. On the other hand, they also reported that the ratio of transactions to standing orders declined (significantly) from 23% in the cancelation treatment to 19% without cancelations. They detected no statistically significant relationship between the two treatments with regard to the limits submitted or regarding the price variance.

2.4.4.8 Other Modifications of Experimental Design

This section contains a number of additional modifications that were explored compared to the original Smith et al. (1988) baseline market. While these treatments do not fit into one of the previous sections in this chapter, they nonetheless offer some interesting glimpses of the factors influencing outcomes in experimental asset markets and were important in the design of the institution chosen for the experimental work.

Williams (1980) reported "on the first series of computer-automated double auction experiments"[72] that aimed to mimic oral double auctions of the type reported in Smith (1962). The computer system he employed (PLATO) "handles all aspects of the experiment except the recruiting of subjects and their payment of

[72] Williams (1980), p. 236.

earnings in cash at the market's conclusion." It accepted inputs via touch screen and seems to have offered similar functionality as current experimental software packages (e.g., z-Tree) for double auction markets. Over the course of his experiments, Williams tested three different rules regarding the acceptance of quotes. The rule still employed in most experiments today (rule 3b in Williams' paper) was that price quotations had to progress so as to reduce the bid-ask spread. Any new bid (offer) had to be higher (lower) than the currently standing best bid (offer). Interestingly, as a second possible institution (3a), Williams (1980) named the rule that whenever a new quote enters the market, it should remain open to acceptance for a number of seconds before it can be replaced by another offer. He wrote:[73]

> "The necessity of having some minimum standing time for each price quote is easily seen if one considers the consequences of a dominant "bumping" strategy where subjects try to rapidly displace the current standing bid or offer with their own. In the absence of a human auctioneer-experimenter to slow things down and maintain order in the market, such behavior would render the act of accepting a particular price quote very difficult. Contract prices might tend to have a high degree of variation as haphazard and panic acceptance occurred."[74]

Finally, Williams' third rule (3c) stipulated that each price quote would be displayed to the market for a minimum number of seconds (as under 3a), but new quotes entered within that minimum display time were queued according to their time of entry and displayed in that order. All participants received continuously updated queue-length information. While an offer was in the queue, its creator could not accept any price quote - he or she was thus blocked from taking any action until the time during which his or her own offer was displayed had expired. Williams expected these opportunity costs to induce participants to refrain from entering new quotations when the queue was long. When reviewing the results of this regime however, he noted that the queues were considerably longer than expected, which he interpreted as a sign of his subjects' fascination with the technology of registering quotes, further documenting the novelty of the computerized trading mechanism at the time of Williams' experiments:[75]

> "It appeared that subjects were deriving sufficient utility from the mechanism itself (using the touch panel to enter price quotes) to offset the costs of queuing. In relation to this it is interesting to note that the number of bids and offers per period in experiment 1 ran about three times the number entered in the oral double auction (approximately 90:30). To the extent that such nonmonetary utility considerations affect individuals' behavior in the market, the experimenter's control on the underlying supply and demand conditions is lessened."

---

[73]Williams (1980), p. 238.

[74]Note that such behavior was not prevalent in the experiment conducted for this book, even though there were cases where subjects reported that a quote was accepted just moments before they themselves clicked the "Accept" button, such that they accepted a quote different from the one they had wanted to accept. Nonetheless, this was a rare occurrence and there was no evidence for any impact on the experimental results.

[75]Williams (1980), p. 245.

A novelty introduced already in Smith et al. (1988) was that they solicited forecasts of next period's mean contract price from their subjects, rewarding the subject with the smallest cumulative absolute forecasting error with a bonus payment of $1. They found that subjects succeed in forecasting prices if they remained approximately constant, exhibited a small trend or followed intrinsic values, while they failed to predict turning points. Experience increased the quality of forecasts.

Ang et al. (1992) used psychological tests to sort subjects according to their respective risk appetites. They reported that in the baseline experiments, their less risk-averse subjects traded at a smaller discount from the more risk-averse subjects. Furthermore they showed that the introduction of a bonus payment in line with the experiments described in Sect. 2.4.4.4 led to a first-period bubble in the market of less risk-averse subjects, but not in the market of more risk-averse subjects. Based on these results, they suggested that excess volatility would be reduced by modifying the regulatory environment so that buyers and sellers face similar costs. King et al. (1993) tried the opposite tack when they introduced significant transaction costs to discourage trading and possibly reduce the occurrence of bubbles. They found that, while mean turnover increased (decreased) for inexperienced (experienced) subjects, mean amplitude and price variance declined.

Gode and Sunder (1993) explored the role of the double auction transaction form by comparing conventional laboratory markets with the results of computer simulated market experiments. They induced supply and demand curves for the single traded good and observed quick convergence to the rational expectations equilibrium in the human subject market. They then ran the same experiment with two types of "zero-intelligence" machine traders, which posted bid and ask quotes randomly. The simulated traders of the first group could only post bids which exceeded their redemption value or ask quotes that were below their cost (zero-intelligence with constraint), while the second group could post any quote within a range of 1–200 currency units, even if they caused them to lose money on the transaction (zero-intelligence unconstrained). This market design permitted the identification of systematic characteristics of human traders by comparing the results from the human subject market with that of the constrained zero-intelligence traders. By comparing the outcome of the constrained zero-intelligence traders with their unconstrained brethren, it also permitted the identification of the effects that ensued from the imposition of budget constraints on a market's traders. The results showed that a progressive narrowing of the opportunity set of the constrained computer traders led them to converge on the rational expectations equilibrium and made their efficiency hardly distinguishable from that of the human agents. While human subjects learned quickly and then stayed at virtually 100% efficiency, the constrained simulations – only through the enforcement of market discipline among unintelligent computer agents quoting random prices – similarly attained an average efficiency rating of 98.7%

Stanley (1994) conducted a market experiment modeled after the Smith et al. (1988) design, modifying the dividend structure, but more importantly, altering the termination rule by introducing uncertainty about the number of periods in the experiment. In his institution, trading lasted for between seven and fifteen periods, with an equal probability of the experiment terminating at the end of any of the periods after the seventh. He found that prices did not converge to fundamental values, but developed much like in previous experiments (i.e., they started out below the fundamental value and increased above it). An exception was that at the end they failed to crash back to the fundamental level, yielding a strongly negative correlation between actual prices and fundamental values. This disconnect between actual and fundamental prices caused Stanley to term this phenomenon a "silly bubble." In Stanley (1997), the author employed the same market structure, but ran three repetitions (rounds) with the same subjects. He reported that – contrary to the usual pattern – bubbles continued to be observed even after the subjects had gained experience and participated in one or two previous rounds. Caution is advised in interpreting these findings, however, since in each article, Stanley only had the financial support to conduct a single session with eight subjects, which is hardly encouraging for the robustness that can be expected of his findings.

Fisher and Kelly (2000) let subjects buy and sell two different assets, trying to gain insights into the relative prices, i.e., the exchange rate between these two assets. Despite observing clear bubbles in the individual asset prices, they reported that the exchange rate converged quickly and then stayed close to its theoretical value. Using forecasts made by their experimental subjects, the authors also found that 22 out of 24 agents acted rationally with regard to the exchange rate, while at the same time participating in markets with significant asset price deviations from their fundamental values. Caginalp et al. (2002) similarly let their traders transact in two different stocks. In ten of their experiments, both stocks were parameterized as value stocks (i.e., with relatively low variance of returns), while in four experiments one stock was a value stock and the other was a growth stock (i.e., had a higher variance). They reported that for their design, the presence of a speculative asset lowered the mean price of the less volatile asset by around 20%, while increasing its variance. This underlines the danger speculative bubbles in some goods pose for the remaining assets in an economy. In a second experiment, they employed a design of two markets in identical assets, where each trader could trade only in one of the two markets, but all traders could observe both markets. Using this institution, they found that increases in the cash endowments of traders in one market lead to increases in the prices in this market, but not in the other.

Smith and Williams (1981) ran 16 experiments with experienced subjects to test the impact on markets of price controls in the form of trading halts triggered by large price movements. They reported that markets with nonbinding price ceilings (floors) near the competitive equilibrium price caused markets to converge to this equilibrium from below (above). They provided evidence that the cause lay in a restriction of the bargaining strategies, predominantly of sellers (buyers). Ackert et al. (2001) also investigated the effect of trading halts on experimental markets. They ran three markets each with a treatment where the market was in continuous operation, with one where large price movements triggered a temporary stop in trading, and with a final set of rules where large price movements triggered a permanent halt of trading for the period. The main difference between their study and Smith and Williams (1981) was that each subject in Ackert et al. (2001) could both buy and sell the asset and that in their design, price limits changed dynamically with the level of the asset price instead of remaining constant as in Smith and Williams (1981). Their results suggested that the market structure employed had not influenced the dissemination of information or the generation of profits, but that trading activity by both informed and uninformed subjects surged prior to a trading halt. They controlled for subjects' current holdings and found that trade was motivated by differing expectations regarding the value of the asset, not by differences in current holdings. Evaluating responses to a questionnaire, Ackert et al. (2001) observed that traders used temporary trading halts to reassess their expectations and strategies. Finally, they documented significantly higher trading volume in the permanent halt regime than in the two other designs. On the one hand they concluded that the so-called circuit breaker rules came with no negative side-effects, but on the other hand they could not document any benefits from this kind of trading halts.

Corgnet et al. (2008) explored the impact of informative and uninformative announcements on bubble characteristics in a Smith et al. (1988)-type market. In the treatment with a message preset by the experimenter, they informed their subjects that a message would be displayed on their screens in periods 3, 7 and 12. This message would say either "THE PRICE IS TOO HIGH" or "THE PRICE IS TOO LOW," and subjects were told that the choice between these two messages would be made by the experimenter before the session started. They conjectured that this design would lend medium credibility to the message, since subjects would assume that the experimenter's choice would be informed. To provide additional insights, they also had a treatment where this message was selected randomly prior to the start of the period (low credibility of the announcement), and one in which it was chosen to correctly reflect the relative difference in prices to fundamental value in the previous period (high credibility of the message). Corgnet et al. (2008) found that – compared to the baseline design without an announcement – the random message design did not significantly affect any of the bubble measures they employed (see Table 7 to 11). They attributed this result to their subjects requiring a necessary minimum level of reliability for a message to have any effect. In the design with messages preset by the experimenter, the "high" ("low") message significantly reduced (did not affect) amplitude and duration of the bubble, as well as the price deviation from fundamental value, for inexperienced subjects (for any subjects). Finally, the message based on actual prices in the market succeeded in significantly reducing the bubble amplitude and a measure of normalized average price deviation.

Noussair and Powell (2008) compare the transaction price process in a market where the fundamental value declines and then increases again (valley treatment) to a market in which the fundamental value of the traded good first increases and then declines (peak treatment). They find evidence for a path-dependency of the bubble phenomenon, in that peak market prices tend to more quickly converge toward fundamental value. This result may be of interest considering that in real markets, investors constantly enter and exit the market. This results in different subjective price histories between investors and could possibly impact market efficiency, particularly if groups of investors (professional investors, naïve investors, etc.) share systematic differences in the timing of their market entry (i.e., the phases of the market cycle in which they are more and less likely to enter and exit the market).

Hussam et al. (2008) combined a baseline treatment of a Smith et al. (1988)-type market with a treatment where – after having run two consecutive baseline rounds with the same cohort – they modified the initial endowment and the dividend structure to see whether this would rekindle a bubble. Their results confirmed that experience significantly reduces bubble amplitude and turnover, but discovered that in the rekindle treatment, the resulting bubble amplitude and turnover (in the rekindle round with twice experienced subjects) are not significantly different from that produced by inexperienced subjects. They concluded that experience is a sufficient condition to eliminate bubbles in static replications of the baseline environment, but not for the changing environment of the rekindle treatment. Conversely, the bubble duration was reduced both in the baseline and in the rekindle treatment. In a more far-reaching result they also reported that in a third treatment that employed the parameterization of the rekindle treatment already in the first period, the bubble amplitude is not reduced even in the third round, while the duration and turnover do decline.

2.4.5 Efficiency in Experimental Asset Markets

> *The ubiquitous tendency for laboratory assets with a well-defined declining fundamental value to trade at prices below this value, then rise above it, and crash near the end of the horizon, has launched experimental inquiries designed to investigate why this is so. [...] Since the participants themselves are mystified by this pattern, interrogating them has not been a source of great insight beyond establishing that they are indeed baffled, much as stock market investors in the economy.*
>
> Porter and Smith (1995)

The most commonly reported dimension of a market's functioning in economics is its informational and allocational efficiency. Naturally, measurements of and reports on market efficiency were provided by a number of experimental studies and also play a prominent role in the presentation of the results in Chap. 4. The following paragraphs review studies from the prior literature which provide evidence on the efficiency of experimental asset markets.

Plot and Sunders (1988) investigated the efficiency of experimental asset markets and showed that, while their experimental markets were fair games and filter rules did not outperform a simple buy-and-hold strategy, they were not efficient in a rational expectations sense.[76] Moreover, even a strategy of trading on the rational expectations equilibrium price would have failed to beat the buy-and-hold strategy,

[76] Cp. Fama (1970) for more on the role of fair games and filter rules in studies on informational market efficiency.

since the markets consistently failed to converge to this price. They concluded that markets that are fair games need not necessarily be efficient. More generally, Sunders (1995) quoted studies showing that the absence of arbitrage opportunities does not imply informational market efficiency. In Gode and Sunder (1994), he and Dhananjay Gode also showed that with regard to the percentage of the available surplus exploited in a double auction market institution, zero-intelligence computer traders were not inferior to human and artificial intelligence traders, as was already mentioned in Sect. 2.4.4.8. In a similar approach (which at that time departed from much of the previous literature) Haruvy and Noussair (2006) showed that a simulated market populated with speculators, feedback traders and passive (fundamentalist) traders generated similar patterns as those they had observed in their experimental markets.

In a more theoretical account, Friedman (1984b) wrote that a generic trader in an experimental double auction market can immediately increase her utility using one of four actions. She can accept the market bid or ask if they – respectively – exceed or fall short of her own valuation, or place a more competitive bid or ask quote if she does not already hold the best quote and the current quotes do not – respectively – exceed or fall short of her valuation. The first two options lead to an immediate increase in utility, while the second two options lead to an increase in the expected value of the trader's position, as long as there is a positive probability that the new bid or ask will be accepted by another trader (which also induces an immediate increase in *utility*, though not necessarily in experimental wealth). Friedman called a trader limiting herself to one of these four actions *myopic*, since the maximization of the expected utility of *final* holdings might also entail accepting an ask (bid) price above (below) her valuation. He also suggested that especially (but not only) inexperienced traders face a tradeoff when faced with favorable market prices. Such a trader may either transact immediately, locking in an expected profit, or hold back in the hope of finding more favorable prices later in the period. Naturally, the option of holding back and waiting declines in attractiveness with the passage of time and the nearing of the end of the period. Friedman (1984b) reported that he occasionally observed a flurry of transactions late in a trading round, which were presumably caused by traders who had waited too long and were then trying to still complete profitable trades in the time remaining before the end of the trading period. Still, according to his findings, experienced subjects seldom missed out on attempted transactions because they had waited too long.

In this context, Friedman coined the term of a no-congestion equilibrium, which he characterized as follows:[77]

> "Roughly speaking, I ask: if the market were unexpectedly held open an extra instant, would anyone definitely wish to change his bid or ask prices, or accept the market bid or ask after all? If not, we have a no-congestion equilibrium."

[77] Friedman (1984b), p. 65.

He then went on to use a simple arbitrage argument to show that in his model three agents are sufficient to yield Pareto optimal final allocations in which the closing market bid and ask prices coincide. He argued that, with only two traders, there exists the possibility of a bilateral monopoly impasse in which both traders look to their counterparty to make price concessions, even in the extra instant. Once a third agent is added to the mix, competition forces prices to the Pareto optimum. In his conclusion, Friedman noted that three main features contribute to the remarkable efficiency of experimental double auction markets. The first is the double auction structure with strictly improving quotes, as it limits a trader's potential impact on prices and conveys high quality information to market participants. This is a marked contrast to the example of a tâtonnement institution, where price quotes are collected by an auctioneer who then announces a market-clearing price. In such a setup, very little information about the distribution of agents' reservation prices is conveyed to the market and there exist extensive possibilities to convey misleading information to the market (e.g., false excess demand). The second characteristic of experimental double auction markets leading to efficient outcomes is the fixed ending time of the trading period, which forces agents to become more myopic if they wish to realize remaining gains from trading. Together with the informational and competitive aspects of the double auction institution, he found that this alone may be sufficient to bring about an efficient final allocation even with agents who initially possess little information about what to expect. Finally, as a third feature that causes allocational efficiency, Friedman named stationary replication, the beneficial effects of which have already been discussed in Sect. 2.4.2 of the present text.

Chapter 3
Experimental Design and Methodology

Another way of dealing with [experimental research] errors is to have friends who are willing to spend the time necessary to carry out a critical examination of the experimental design beforehand and the results after the experiments have been completed. An even better way is to have an enemy. An enemy is willing to devote a vast amount of time and brain power to ferreting out errors both large and small, and this without any compensation. The trouble is that really capable enemies are scarce; most of them are only ordinary. Another trouble with enemies is that they sometimes develop into friends and lose a good deal of their zeal. It was in this way that the writer lost his three best enemies.

György von Békésy (1960)

Due to the number of different experimental market designs and treatments, different articles in the literature sometimes use the same terms to refer to different components of an experiment. As an example, the term of an "experiment" can refer to one run of an experimental market, to the set of runs conducted with the same subjects or treatment design, or to the set of all runs and treatments conducted to answer a research question (depending on the study one chooses from the literature). To ensure that the meaning of the terms used in the description of the design and in the presentation of the results are unambiguous, the following chapters require some conventions with regard to the nomenclature employed. Such a convention is specified in the following paragraphs:

In the experimental work conducted for and presented in the following chapters, a *period* is the time of continuous trading in the experimental market between two

S. Palan, *Bubbles and Crashes in Experimental Asset Markets*,
Lecture Notes in Economics and Mathematical Systems 626,
DOI: 10.1007/978-3-642-02147-3_3, © Springer-Verlag Berlin Heidelberg 2009

consecutive dividend payments.[1,2] In the experiments reported below, a period lasted between 240 and 300 s (depending on the experimental treatment, as will be described later in this chapter). After each period, subjects were presented with the period-end screen, which gave them information on stock trading, option trading and the evolution of their portfolio over the period, as well as an outlook on future periods. A period can be thought of as loosely corresponding to a trading day in real-world financial markets.

Fifteen periods formed a *round* in the experiments reported below. Between periods, both stock and cash endowments were carried forward; between rounds, they were re-initialized. One group of subjects participated in either two or three rounds. In the literature, "market," "session," or "experiment" are sometimes used as synonyms for a round.

Rounds can be aggregated to experimental *sessions*, a term that refers to a set of rounds conducted with the same group of subjects and at the same calendar date.[3] Within a session, the only design difference between the rounds is subject experience and initial endowments. The group of subjects participating in one session is referred to as a *cohort*.

Every session[4] follows a *treatment* or *institution* – a set of procedures and parameters that forms a specific experimental design. In the case of this work, the two treatments employed differed with regard to the digital options' maturity dates.

Finally, the set of sessions (belonging to one or more treatments) which are run with the aim of answering a specific research question is referred to as an *experiment*. This book reports on one experiment. This experiment was designed to explore the question of whether providing subjects with the opportunity of trading in a digital option market improves the informational efficiency of the underlying spot asset market.

[1]The first period is the time of continuous trading between the start of the experiment and the first dividend payment.

[2]In some Smith et al. (1988)-type experiments, dividends were not paid at the end of each period. In that case a period could be defined as the time of continuous, uninterrupted trading from a market's open until its subsequent close. This definition is more general but less clear, which is the reason why it was not chosen for the work presented here.
Nonetheless, in most experiments the reader will have no problem recognizing a period for what it is, since the word is used similarly in most studies, unlike the terms "session", "round", or "experiment", which can have different meanings coming from different authors.

[3]As mentioned before, other studies sometimes conducted repeat rounds with the same subjects on different days.

[4]In other experiments, a session may contain rounds following heterogeneous treatments. In the work reported here, every round within a session followed the same treatment.

3.1 Interface Design

The heart of the experiment consisted of an electronic double auction market programmed and conducted with the software z-Tree, version 3.2.11.[5] Generally, the design of the trading screen, the information displayed and the mechanics of interaction with the computer and with other subjects is a crucial part of any computerized experiment. In the experimental sessions conducted for the research project reported in this book, it gained particular importance due to the high complexity of possible transactions and the large amount of information presented to subjects. While in most previous studies, subjects had traded in one market, they could trade in both the stock and the option market in the present experiment. This fact not only increased the number of possible courses of action, it also strongly amplified the sensory input subjects were exposed to. At any time, they had to monitor changing price quotes in two markets, which entailed processing the implications of new developments within one market not only for that market,

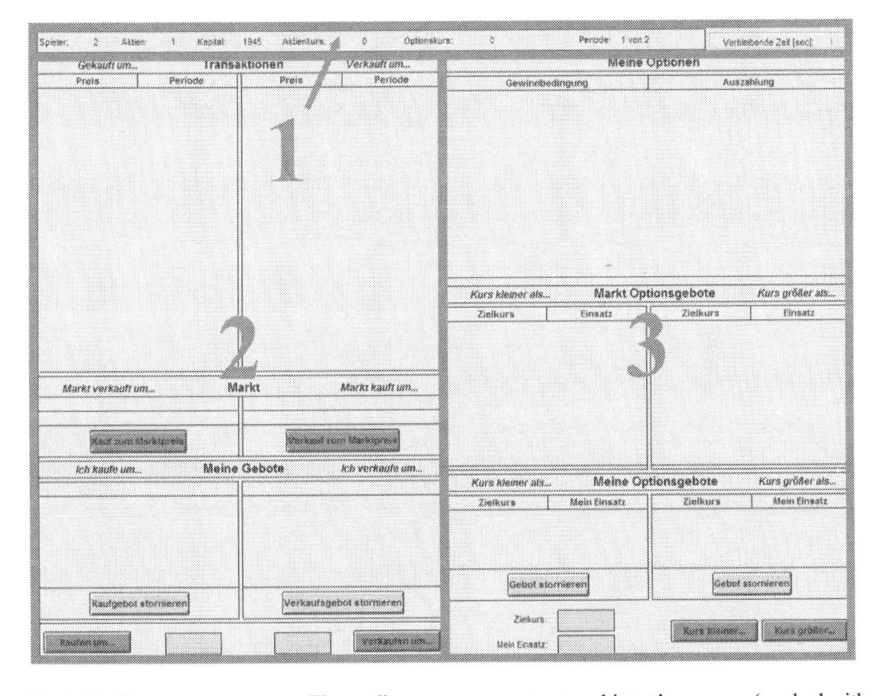

Fig. 4 Trading screen structure. The trading screen was structured into three areas (marked with numbers and framed): Area 1 at the top of the screen is the information area, area 2 to the left is the stock market, and area 3 to the right is the option market

[5] Cp. Fischbacher (2007).

but also for the second market. A new stock quote, for example, not only possibly contained information regarding future stock prices, but also created a new informational situation for option quotes and for the value of existing options held by the subject. As De Jong et al. (2006) so succinctly put it: "A barrier to experimental research that incorporates options is the necessary complexity of such markets."[6,7]

The layout of the trading screen was designed with the goal to provide a clear structure, which would supply the subjects with the information and interaction possibilities necessary, while minimizing the sensory input and the possibility for confusion as far as possible. As Fig. 4 shows, the trading screen was structured into three areas. Area 1 at the top of the screen contained information on the subject's current portfolio of stock and experimental currency, and listed the last stock and option transaction prices.[8] It also displayed the current period number and the time remaining until the end of the period (in seconds). Area 2 to the left of the screen was the stock market. It gave subjects the possibility to create new bid and ask quotes, or to accept the best currently outstanding bid or ask quote. In addition, it also informed them about their personal stock transactions to date in the form of a two-tuple of period and price (over all previous periods within the current round). Furthermore, it displayed the current best bid and ask in the stock market, the subject's own outstanding stock quotes, and all outstanding option quotes. A stock quote consisted only of the price, since all quotes were limit orders with the quantity fixed at one share of stock. Area 3 to the right of the screen was the option market. It listed the options currently held by the subject, displayed all option quotes currently open in the market, and gave subjects the possibility to post new option quotes themselves. An option quote consisted of the winning condition (i.e., "The stock price at the end of period 8 is larger than 100.") and the amount invested by the subject posting the quote, i.e., the "option volume," while an option contract was listed as a two-tuple of the winning condition and the payoff in the case that the option would be in the money at its maturity.

In addition to the information displayed on the trading screen, subjects' information sets during the trading periods included the number and identity of traders,[9] but subjects received no information allowing them to map trader identities on actions in the experimental market, nor on the individual performance of subjects other than themselves. At the end of each period, subjects were informed about changes in their portfolio due to stock transactions over the period, about their gains

[6]De Jong et al. (2006), p. 2247.

[7]The complexity of the experiment may be illustrated by noting that the z-tree code ran to more than 1,600 lines and took about 700 h to write and test.

[8]In the case of options, the price displayed was the strike price of the last option quote accepted by a subject in the market.

[9]Providing this information was unavoidable, because due to lab space constraints all subjects were in the same room for the duration of the experimental rounds.

and losses as well as about capital frozen due to option quotes and holdings.[10] Furthermore, they received information about the dividend draw and its impact on their portfolio, about the remaining minimum, expected, and maximum dividend value of one share of stock, and about the expected terminal value of their current portfolio. In order to enable subjects to choose their actions optimally in both markets, a voluminous set of additional information was provided in the instructions.[11]

In designing the interactions with the market software, great care was taken to strive for simplicity and to prevent the maximum possible number of errors. As one such measure, subjects could enter bid (ask) quotes for shares of stock using the "Buy for..." ("Sell for...") button as long as their price was not higher (lower) than the best[12] currently outstanding ask (bid). If a subject for example entered a bid that exceeded the best outstanding ask quote, a message window with the following information was displayed: "A seller is willing to accept the same or a lower price. Instead of entering a purchase order, please accept the best outstanding sales order!" To make an immediate stock purchase (sale), subjects had to press the button labeled "Buy at market price" ("Sell at market price"). The rationale behind separating the buttons necessary to submit limit orders from those for posting market orders was to limit the number of inadvertent quasi market orders. If a subject for example wished to enter a bid quote with the price of 100 and mistyped the price as 1,000, this order can be thought of as a quasi market order, in that its price would probably exceed all outstanding asks and lead to an immediate transaction. The separation of the two types of buttons prevented this kind of error. Similarly, if a subject entered a bid quote that exceeded one of her own ask quotes, the error message read: "This order conflicts with one of your sales orders. Please check your entry or cancel the conflicting sales order!" In this way, subjects were made aware of inconsistencies in their actions.[13]

---

[10] When subjects entered into an option contract, the amount invested into the contract was deducted from their cash accounts (i.e., frozen), forming the equivalent of a margin of 100%. The option market design is explained in more detail in Sect. 3.3.

[11] The original (German) instructions are available online from the publisher. They can be downloaded from: http://www.springer.com/9783642021466.

[12] As will be made clear in Sect. 3.2, every quote in the stock market was for one share of stock, and no two quotes outstanding in the market at the same time could have the same price. Therefore, bid and ask quotes could easily be sorted by price, thus yielding a unique best (i.e., highest) bid and a unique best (i.e., lowest) ask quote.

[13] One could argue that investors in real markets would not be prevented from entering such orders, and that such behavior could theoretically constitute part of the reason for the observation of bubbles in asset markets, which is the phenomenon the experiment was designed to research. One counterargument is that in real markets, brokers and intermediaries can be assumed to act as a filter by sometimes making investors aware of such inconsistencies. Nonetheless, in the end the use of this safeguard measure in the experiment comes down to a judgment call, where the benefits from preventing mistakes that were unique to the experimental setting and the student subject pool were considered to outweigh the caveat of a design deviating slightly more from real-world markets.

In the option market, similar safety mechanisms were instituted. Subjects could not submit option quotes that conflicted with other option quotes they had outstanding. In the case that a newly-entered option quote immediately led to a transaction, subjects were asked: "This order leads to the creation of an option contract. Do you want to proceed?" Subjects were thus made aware that their new option quote interacted with existing outstanding quotes by other subjects in such a way that it would lead to the immediate creation of an option contract, and could decide to proceed or revise their entry.

Safety measures were also programmed to prevent subjects from going short in cash or stock, or from entering a position that would lead to a short position later on. For example, if a subject entered into an option contract, the amount invested was immediately deducted from her account. Similarly, if a stock or option transaction took place, all other unmatched stock and option quotes of the two subjects involved were checked for feasibility. Bids in the stock market that the subject could no longer fulfill were automatically deleted; asks were deleted as soon as a subject had no more shares to sell; and the amount invested in option quotes was adjusted to the greater of the amount invested into that quote prior to the last transaction and the subject's cash holding after the last transaction. In this way, at no time were any quotes in the market which could not be funded or delivered upon by the subject holding them.

3.2 Stock Market

By virtue of exchange, one man's prosperity is beneficial to all others.

Frédéric Bastiat, 1801–1850

In the experimental markets conducted for this study, shares of the single stock could be traded at any time during the 15 periods and by any of the between 11 and 14 subjects in both treatments. Each stock transaction led to the share of stock being transferred from the seller's to the buyer's inventory and cash matching the transaction price being transferred from the buyer to the seller. As shown in Table 3 earlier, each subject was initially endowed with between one and three shares of stock. The trading mechanism was a classical computerized continuous double auction with a closed order book displaying only the single best bid and ask prices.[14] A feature which this study's experimental stock market shared with many similar experimental markets (e.g., Campbell et al. 1991) was that bids and asks had to be improving: Any new offer to buy (sell) a share had to have a higher (lower) price than the current best

[14]In the first experimental session, the order book was open, displaying all outstanding bid and ask prices. This increased the amount of information displayed on the screen (i.e., the sensory input), without appreciable effects on subjects' behavior. To reduce the sensory load on the participating agents, it was changed for all subsequent experiments, leading to no observed difference in subjects' actions and no noticeable impact on results. It should nonetheless be mentioned here, since effects from this change in market presentation – though not apparent – cannot be ruled out with certainty.

standing bid (ask). Subjects could thus enter new bids (asks) if they improved the current highest (lowest) outstanding offer to buy (sell). This improvement rule is also common in real markets – it is for example part of the regulations governing trade at the NYSE.[15] Finally, subjects could also buy (sell) a share at the prevailing ask (bid) price, i.e., submit a market order, by clicking a button labeled “Buy at market price” (“Sell at market price”) – as was already briefly mentioned in the previous section. Traders could submit and cancel quotes in the stock and option markets at any time, subject to no-short-sale and no-margin-buying constraints.[16] Furthermore, feasible prices were constrained to integer values in the interval [1, 10,000] (in cents).

At the end of each period, each share of stock paid a random dividend, drawn from a discrete uniform distribution with four possible dividend values (0, 8, 28 or 60 Euro cents). The stock had no terminal value, causing its fundamental value to decline by the expected dividend (24 cents) in each period, as illustrated by the stepwise function plotted in the solid line in Fig. 3 on page 7. The fundamental dividend value and the possible payoffs from stock were known to all subjects.

Note that posting an offer in the stock market (as well as in the option market) amounted to writing what O’Hara (1995) described as a free option. By submitting a limit order, the trader committed to trade at a particular price, giving the other market participants the choice to conduct a transaction at that price at any time prior to either the cancelation of the limit order through the original trader or to its expiry at end of the period.[17] O’Hara (1995) and Easley et al. (1996) pointed out that, due to the changing nature of market prices and expectations, such an option exposes its writer to the risk of the option being exercised at a point in time unfavorable to its creator. This risk can be mitigated by constant monitoring, but this kind of monitoring similarly imposes costs on the option writer. In real-world financial markets, this argument suggests that the cost of this free option should reduce the posting of limit orders and increase the spread, or lead to limit orders containing prices which are farther away from the current best bid and ask than would otherwise be expected. Considering the large number of transactions in the experimental sessions reported here, this potential problem seems to be of limited importance in this particular setting. One reason for that finding might be that, for the duration of the experiment, subjects could not leave their computer screen, implying that the marginal cost of monitoring the market were relatively small.

3.3 Digital Option Market

In the experimental design used in both reported treatments, subjects had the opportunity to trade not only in a stock market, but also in a market for digital

[15] See NYSE (2008), rules 70 and 71.

[16] No margin buying in this context meant no margin buying over and above a loan of €10.00 granted each subject with each round’s initial endowment.

[17] Cp. O’Hara (1995), p. 197.

options. The instrument of the digital option was chosen for the following two reasons: First, it was the aim of this research effort to investigate what effect online betting sites and prediction markets that focus on the prediction of financial market prices have on the efficiency of these financial markets. In most of these new online trading places, the instrument being traded is either a futures contract or a digital option. The second type was chosen here partly because many bets in all kinds of contexts can be viewed as being digital options, since they bear zero initial cost (neglecting possible margin requirements) and promise a fixed gain (loss) if the bettor wins (loses). Second, the digital option was chosen because it offers a fixed payoff, conditional on the option being in the money at maturity. This payoff pattern is easier to understand than the more complex variable payoff profile of a standard option.[18] Since the market environment in the experiment was relatively complex, especially considering the subject pool of mainly bachelor and master students, simplification was an important consideration.

The option market described here was loosely modeled after the transaction form employed on redmonitor.com. Part of this trading mechanism is redmonitor's use of digital options where traders can set their own strike prices. Options in their market are formed once an outstanding option quote finds a counterparty. This transaction form was also used in the experiment reported here. Subjects could specify a price they expected the future stock price to exceed or fall short of (long or short position,

Treatment DO8

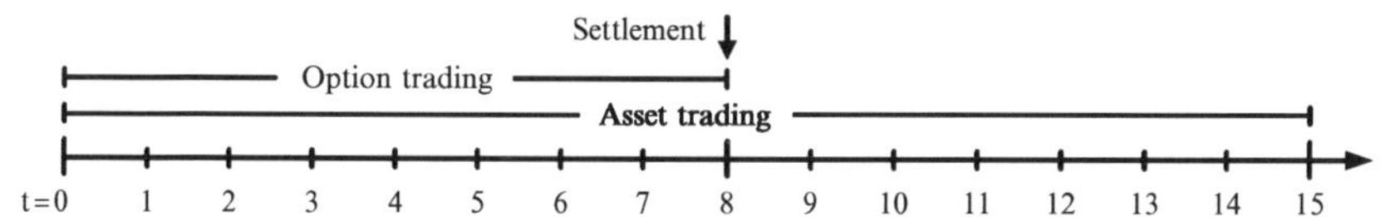

Treatment DO5/10/15

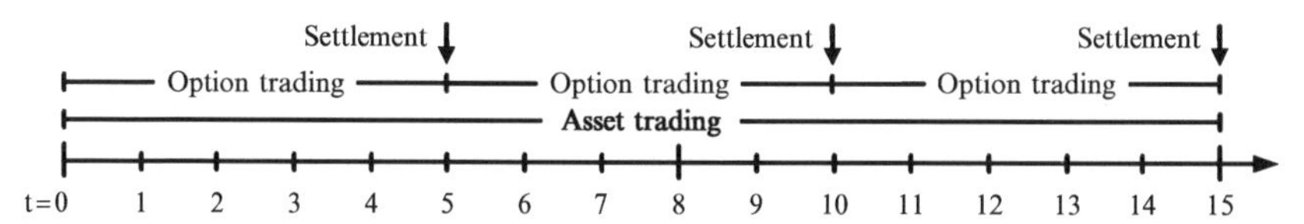

Fig. 5 Time structure of spot and option trading. Illustration of the time structure of spot and option trading in the DO8 (upper panel) and DO5/10/15 treatments. The stock trading structure is identical between treatments; option trading differs with regard to option maturity and because no option can be traded in periods 9–15 of treatment DO8

[18] This statement may seem dubious to somebody who has tried to price a digital option. However, while digital options are difficult to price analytically, subjects typically do not perceive the need to do so. Risk-neutral, rational decision-makers enter into a digital option contract if the probability of success (the expected probability of the option being in the money at maturity) exceeds 0.5, a point that is derived later on in this section. Similarly, most subjects intuitively grasp that a digital option is "attractive" if they are "more likely than not" to "win" when entering into it.

which is equal to holding a call or a put in the case of digital options) and also specify the amount of their wealth they wanted to invest in this option quote. As described in more detail in Sect. 3.4, the two treatments differed with regard to the maturity of the option(s). In the digital options treatment DO8, options could be traded from periods 1 to 8, all options matured at the end of period 8, and were judged using the stock price at the end of period 8. Starting from period 9, the option market was closed. In DO5/10/15, options that were created in periods 1 through 5 matured at the end of period five, options created in periods 6 through 10 matured at the end of period 10, and options created in periods 11 through 15 matured at the end of period 15. This time structure is illustrated in Fig. 5 above.

At the options' maturity, they were settled based on the price of the last stock transaction before the end of the settlement period. A subject's payoff from an option at maturity M, $PO_{t=M,\theta}$, was:

$$PO_{t=M,\theta} = \begin{cases} 2 \cdot SI \cdot (1 - \theta) & \text{if } S_M < X \\ SI & \text{if } S_M = X \\ 2 \cdot SI \cdot \theta & \text{if } S_M > X \end{cases}$$

where $M \in \mathbb{M}^{DO8} = \{8\}$ is the option's maturity date in the DO8 treatment, and $M \in \mathbb{M}^{DO5/10/15} = \{5, 10, 15\}$ in the DO5/10/15 treatment, θ is a binary variable equaling unity if the subject holds a digital call and zero if the subject holds a digital put option, SI is the stake invested into the option by the subject (equaling that invested by her counterparty), S_M is the stock price at maturity and X is the option's strike price. The *profit* from such an option then is $PO_{t=M,\theta} - SI$.[19]

The hypothesis explored with treatment DO8 was that throughout the first eight periods, the constant visibility of the expected stock price in period eight (as revealed in the digital options market and displayed at the top of each trader's screen) would lead to a reduction in the extent of the observed bubble. To this end, subjects were specifically made aware of the expectation that option market prices would contain information about market participants' expectations regarding the stock price in the future. This design is related to that of Porter and Smith (1995), summarized in Sect. 2.4.3.2, who found that the existence of a futures market complementing the stock market led to a significant reduction in bubble amplitude for all, and of turnover for experienced subjects. In their design, the futures market allowed traders to express their expectations regarding the future price by quoting limit prices and specifying whether they desired to buy or sell (e.g., "I want to buy a share of period-eight stock at a price of no more than 100"). In the digital option design, subjects could, in addition to the information about price and

[19] Since one period in the experimental market can be thought of as corresponding to one day in a real market, no interest was paid on cash holdings, nor was it taken into account in any value calculations. This is reasonable also because one round lasts around 2 h in real time, making virtually zero any possible real-world interest requirement founded on a time-preference argument.

direction, reveal the strength of their convictions through the amount of money they invested in their digital option quotes (e.g., "I bet 300 cents that the stock price in period eight will exceed 100"). The hypothesis tested here was that this more direct revelation (and backing up) of expectations would lead to a similar or larger reduction in bubble indicators than the Porter and Smith (1995) futures design. The setup chosen in treatment DO8 – in everything but the choice of derivative instrument – conformed closely to the design of Porter and Smith (1995). This was a deliberate decision to ensure the comparability of the results of this study with Porter and Smith's conclusions and with the outcomes of other studies using a similar experimental design.

The treatment design DO5/10/15 was introduced later, when the first sessions run under the DO8 institution exhibited a persistent structural break in market prices between the eighth and ninth period (particularly in the rounds with experienced subjects), a point in time that coincides with the option maturity date in this design. The experiments conveyed the impression that prior to the end of period 8, traders' attention with regard to the stock price was focused on the levels of their option strike prices. Only once the option outcomes were decided did they seem to let the fundamental dividend value reenter their stock price expectations formation process. An example of this pattern is illustrated in Fig. 6 below. The DO5/10/15 treatment was specifically designed to test whether more frequent option maturity dates would

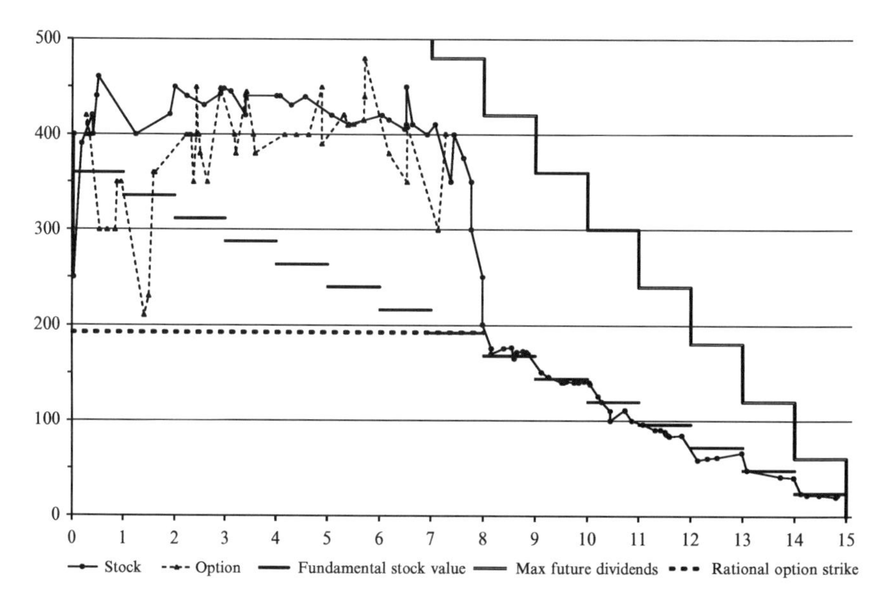

Fig. 6 Example price plot showing the structural break in the DO8 treatment. The figure plots the stock price (solid line with circles), option price (broken line with triangles), fundamental stock value (stepwise decreasing function, single solid line), rational option strike price (dotted line) and maximum value of future dividends (stepwise decreasing function, double solid line) in the second round of session 2. The structural break in period 8 – from prices increasing relative to the fundamental value, to prices closely tracking the fundamental value – is clearly visible

hasten the return to fundamental values and eliminate or ameliorate the effect of the option maturity date acting as an attractor.

In the option market, subjects could submit new option quotes, with their investment bounded from above by their cash holdings. Just like in the stock market, feasible prices were constrained to integer values in the interval [1, 10,000] (in cents). Immediately upon entry, new quotes were checked against existing quotes to determine whether the new quote created any contradictions with existing quotes. If the new quote contradicted an outstanding old quote, the two quotes were matched and converted into option contracts, with any remaining partial quotes being entered into the order book.

An example of the processing of quotes and their conversion into option contracts can be illustrated as follows:

Example 1. Processing of New Option Quotes Without Conflict

(1) Subject A submits a call option quote with a strike price of 101 and an invested amount of 450 cents (i.e., "I bet 450 cents that the stock price in period eight will exceed 101.")
(2) Subject B submits a call option quote with a strike price of 103 and an invested amount of 200 cents (i.e., "I bet 200 cents that the stock price in period eight will exceed 103.")
(3) It is determined that there is no conflict between these two quotes. (The stock price at the end of period eight can be – at the same time – both higher than 103 and higher than 101.)
(4) Both option quotes are displayed on subjects' screens, waiting for matching with new quotes that might be in conflict with them.

In this Example 1, the two option quotes are compatible and no option contracts are being created. The following Example 2 continues from the above exposition:

Example 2. Processing of New, Conflicting Option Quotes

(1) Subject C submits a put option quote with a strike price of 98 and an invested amount of 250 (i.e., "I bet 250 cents that the stock price in period eight will be lower than 98.")
(2) It is determined that there is a conflict between this quote and an existing quote. (The stock price at the end of period eight cannot be both higher than 101 and/ or higher than 103, and lower than 98.) For the new quote, a matching with the quote of subject B is most favorable, since subject C contracted for a digital put option and the probability of the stock price being lower than 103 in period eight is higher than the probability of it being below 101.
(3) An option with a strike price of 103 and an invested amount of two times 200 cents (the minimum of the two invested amounts) is being created. Subject B's quote has thereby been completely transformed into an option contract.
(4) Of subject C's new quote, an invested amount of 250 – 200 = 50 cents is left over. This quote conflicts with the option quote of subject A.

(5) An option with a strike price of 101 and an invested amount of two times 50 cents is being created. Subject C's quote has thereby been completely transformed into an option contract.
(6) Subject A's quote is modified by setting the new invested amount to $450 - 50 = 400$. It remains displayed on subjects' screens, waiting for matching with other quotes that might be in conflict with it.

In Example 2, a conflict occurred between the new and the existing option quotes, which was then processed by a transformation of option quotes into option contracts until all conflicts were resolved. In this process, the option contracts are created in such a way that they optimize the position of the subject submitting the most recent quote. The reasoning behind this rule is that the subject submitting the last quote (subject C in the above examples) could have achieved the same result by submitting first a put quote with a strike of 103 and an invested amount of 200, and then another put quote with a strike of 101 with an invested amount of 50 cents. In the experimental sessions, digital option market quotes that had bet on prices lower (higher) than the specified strike price were being matched with the contradicting call quote with the highest (lowest) price first in both treatments. Nonetheless, the experimental digital option market was not subject to an improvement rule like the stock market. It did, however, follow a rule of time priority, ensuring that quotes entered first were executed first in the case of equal strike prices.

Generally, when a subject posted a new option quote, nothing happened if the new quote did not contradict any outstanding quotes. If there was a contradiction, a new option was created and a cash amount corresponding to each counterparty's investment in the newly-formed option was deducted from their inventories, acting as a margin of 100%. At the option's maturity date, its outcome was then decided and the sum of the investments paid to the subject holding the side of the option that was in the money. In the case where the option matured at the money, each counterparty's investment was returned.[20] Option quotes that had found a counterparty and had therefore been transformed into a binding option contract were uncancelable. However, subjects were free to post a new option quote which, if accepted, would effectively close their position.[21]

A minimum rationality condition for trader s posting a call option offer can be derived as follows. First off, the expected profit to trader s from the call option is

$$\mathrm{E}_s[\Pi_s] = Pr_s(p_{t=M} < X) \cdot (-SI) + Pr_s(p_{t=M} = X) \cdot 0 + Pr_s(p_{t=M} > X) \cdot SI$$

[20] This happened in 0.8542% of all cases. More specifically, 0.7407% of all option contracts in the DO8 treatment and 1.0243% of the DO5/10/15 treatment options were found to be at the money at maturity, which corresponds to approximately one occurrence per experimental session or 0.5 occurrences per round. Due to the low frequency of this event, different payment regimes for this case were not explored.

[21] See Crowley and Sade (2004) for a study analyzing the effect of permitting the cancelation of orders on trading volume and prices in a double auction environment.

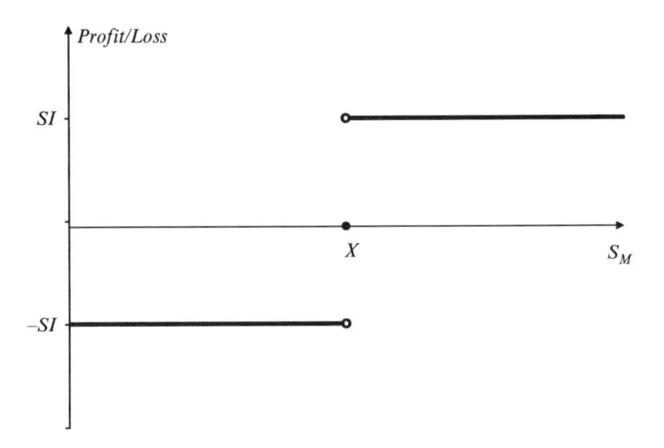

Fig. 7 Digital call option profit at maturity. This figure plots the profit (payoff minus initial investment) from a digital call option as a function of the price of the underlying at the option maturity date

where $\mathrm{E}_s[\cdot]$ is the subjective expectations operator of subject s, Π_s is her profit from the option, $Pr_s(\cdot)$ is the subjective probability operator, $p_{t=M}$ is the stock price at the option's maturity date, and X is the option's strike price, with the other symbols as defined before. Note that this follows from the fact that the initial investment into the option by each subject equals SI, and that the total payoff from the option at maturity is $2 \cdot SI$, the sum of the individual investments (see Fig. 7 for an illustration).

For a trader who is not risk-loving, $\mathrm{E}_s[\Pi_s] > 0$ is a necessary condition to enter into a digital call option contract. From this it follows that

$$Pr_s(p_{t=M} < X) \cdot (-SI) + Pr_s(p_{t=M} = X) \cdot 0 + Pr_s(p_{t=M} > X) \cdot SI > 0$$

$$\Rightarrow Pr_s(p_{t=M} > X) \cdot SI > Pr_s(p_{t=M} < X) \cdot SI$$

$$\Leftrightarrow Pr_s(p_{t=M} > X) > Pr_s(p_{t=M} < X) \tag{1}$$

Equation (1) says that, for a rational risk-averse or risk-neutral trader to post a call option offer, she must expect the future stock price to lie above her strike price with a probability greater than her perception of the probability of the stock price lying below her strike. A similar argument shows that the condition for such an individual to enter into a digital put option contract is that the subjectively perceived probability of the stock price at maturity being smaller than the strike price needs to be greater than her estimate of the probability that the stock price at maturity will be larger than the strike price. Note that if all subjects have homogeneous expectations (and all subjects are rational maximizers of expected utility, where utility is a function only of the final payoff), this argument implies an equilibrium of no trade (see Sect. 2.4.1.5 for a discussion of no-trade equilibria).

In addition to the above considerations, traders must also consider the probability of obtaining an option whenever they post an option offer that they can see from the order book does not immediately lead to a transaction. A trader posting a call option therefore has the incentive to quote as low a strike price as possible, but must balance the gain from quoting a lower strike price (and thereby obtaining an increased probability of the option expiring in the money) against the decreased probability of making a trade. The reverse is true for a prospective put option writer.

3.4 Description of the Experimental Sessions

3.4.1 Subject Pool

The subjects were students recruited in bachelor and master courses in banking, finance and economics.[22] To be more precise, students were informed about the experiment in these courses and could voluntarily write their names on a participation form. Subjects were in no way coerced to submit their names, as is evidenced by the large majority of students approached who did not decide to participate. The experiment was conducted outside of classes and subject performance was in no way connected to the evaluation in the courses. Eckel and Grossman (2000) found that pseudo-volunteer subjects in experiments conducted during class-time behaved in a more extreme manner than conventionally recruited subjects – everything else being equal. This possible source of bias and noise was excluded from the experiments conducted for this book by using only real volunteers and conducting the experiments outside of class-time.

Table 1 lists statistics on the subject pool, grouped by session and treatment. Apart from demographic data, the table also contains information on subjects' previous experience with laboratory experiments and with trading in financial markets, as well as a self-appraisal of their knowledge in the area of finance and of their understanding of the instructions provided.

[22] In Graz, the bachelor-level courses which subjects were recruited from included: Financial Instruments and Financial Markets, Banking & Credit Management, Treasury and Cash Management, and Fiscal and Economic Policy. The master-level courses were: Corporate Finance, Bond Pricing, Stock Pricing, and Option Pricing. In Klagenfurt, the single bachelor-level course was Investment Banking, and the single master-level course was Asset Management. Some bachelor, master, and also a few doctoral students were also recruited outside of courses by word of mouth. The subject pool in session 1 is an exception, as it was conducted using members of the faculty as subjects. Following this first session, minor changes in the instructions and in the screen layout of the trading platform were made. The results of the later sessions did not differ markedly from those of the first. This can be interpreted as a sign of robustness and was the reason that the first session was included in the final analysis.

Table 1 Subject statistics

Measure	DO8					DO5/10/15				Both
Session:	1	2	3	4	All	1	2	3	All	All
Number of subjects	12	12	11	12	11.8	13	12	14	13	12.3
Age	29.7	24.4	24	22.7	25.7	23.2	25.9	24	24.4	25.1
Female[a]	58.3	33.3	45.5	3.33	44.3	0.0	41.7	42.9	28.2	37.5
Bachelor's degree[a]	8.3	25.0	63.6	33.3	29.2	30.8	8.3	14.3	17.9	24.5
Master's degree[a]	50.0	16.7	18.2	8.3	26.4	0.0	16.7	0.0	5.1	17.4
PhD[a]	33.3	0.0	0.0	0.0	11.3	0.0	8.3	0.0	2.6	7.6
Had previously traded stock[a]	58.3	41.7	63.6	33.3	50.0	38.5	50.0	35.7	41.0	46.2
Had previously traded options[a]	16.7	16.7	18.2	0.0	13.2	0.0	16.7	7.1	7.7	10.9
Participated in ≥ 1 prior experiment[a]	66.7	8.3	18.2	41.7	34.0	7.7	33.3	7.1	15.4	25.6
Participated in ≥ 1 prior security trading experiment[a]	25.0	8.3	16.7	33.3	21.3	41.7	16.7	8.3	7.7	15.1
Finance knowledge[b]	2.11	2.17	2.59	2.08	2.22	2.19	2.04	2.04	2.09	2.16
Understanding of instructions[b]	2.61	2.50	2.82	2.75	2.66	2.65	2.63	2.75	2.68	2.67

[a]Percent of all subjects
[b]Mean answer on a scale of 0–4, with 0 indicating no financial knowledge or understanding of the instructions, and 4 indicating the maximum possible understanding
Statistical properties of the subject pool, means by session and treatment. All values were solicited by computerized questionnaires answered by subjects after each round. The table contains information by session on the number of subjects, on their age, sex, highest academic degree, previous experience in trading stocks and options, previous experience in laboratory experiments in general and in experiments involving the trading of securities in a market setting, subjective perception of their knowledge in the area of finance, and subjective perception of how well the subject had understood the instructions

The difference in the subject pool of the first session is reflected in the age and education categories, as well as in the prior experience in economics experiments. Interestingly, it is not apparent in the answers on question 17 of the questionnaire, inquiring about traders' knowledge of finance and capital markets, which was not very reliable from an objective point of view. Several bachelor students chose the reply corresponding to the highest possible finance knowledge, while several members of the faculty chose the lowest. The possible interpretation could be that subjects reported there how they judged their knowledge in comparison to people stemming from a similar demographic and educational group, i.e., the specific subjects who participated in the experimental session with them. Such a mechanism would explain the observed numbers, in that the average faculty member believed to be as knowledgeable in the area of finance, compared to her fellow faculty members, as the average student believed herself to be, compared to her fellow students.

3.4.2 Session Layout

Table 2 gives an overview of the seven experimental sessions. The first five sessions were conducted at the Karl-Franzens-University Graz, the last two at the Alpen-Adria-University Klagenfurt. The reason for the change in location was that the pool of possible subjects had been temporarily exhausted in Graz. Conforming with the official language at both universities, all experiments were held in German, using a native German speaker as the experimenter and using German as the language of both the instructions and the language of the computer program. Session 1 comprised three rounds; all other sessions comprised two rounds. All sessions were conducted in university computer labs, overseen by the author as the sole experimenter.

A frequent point of criticism leveled at experimental studies is that too many variables are changed between treatments, such that the effect of the modification of individual treatment parameters cannot be isolated. This problem was avoided in the present study by designing two treatments which differed solely with regard to the option maturity dates. In DO8, option trading was possible from periods 1 to 8, with all options being settled at the end of period 8. In DO5/10/15, there was option trading from period 1 to 5, from 6 to 10 and from 11 to 15, with options being settled at the end of the last trading period in each interval. The introduction of the second treatment design was motivated by results from the first four sessions, a point that will be dealt with in more detail in the discussion of the results in Sect. 4.2.3. See Fig. 5 in Sect. 3.3 for a graphical representation of spot and option trading under the two treatments.

The reason that the experiment was limited to two option treatments and did not include a baseline treatment for comparison purposes lies in the large number of previous studies employing a Smith et al. (1988)-type design, which provide a broad benchmark for the comparison of results. This is of essential importance when considering the dictate of economically efficient research. The experiments conducted over the course of this research project encompassed 86 subjects, participating over a total of more than 500 person hours, and receiving an overall

Table 2 Session layout

Session	Treatment	Date	Location	Number of subjects	Rounds
1	DO8	2007-11-10	Graz	12	3
2	DO8	2007-12-01	Graz	12	2
3	DO8	2007-12-17	Graz	11	2
4	DO8	2007-12-18	Graz	12	2
5	DO5/10/15	2007-12-19	Graz	13	2
6	DO5/10/15	2008-02-13	Klagenfurt	12	2
7	DO5/10/15	2008-02-14	Klagenfurt	14	2

Five experiments were conducted at the Karl-Franzens-University Graz, Austria, and two at the Alpen-Adria-University Klagenfurt, Austria. All experiments were conducted by the author as the sole experimenter. In the DO8 treatment, option trading was possible from periods 1 to 8, with all options being settled at the end of period 8. In DO5/10/15, there was option trading from period 1 to 5, from 6 to 10, and from 11 to 15, with options being settled at the end of the last trading period in each interval. Session 1 comprised three rounds, all other sessions comprised two rounds

compensation of € 2,450.40. Moreover, students had to be recruited from two universities in order to be able to populate the necessary experimental sessions in the given timeframe. Due to the high resource requirements of this type of analysis it becomes clear that it would have been impractical to conduct further experiments to establish baseline results for the purpose of having a base for comparison, considering that such a benchmark already existed in the economic literature.

A session was structured as follows: The subjects arrived and were seated at computers. The instructions were handed out and read by the subjects (approximately 1 h). The experimenter took special care to make sure that subjects understood the evolution of the stock's fundamental value. The instructions were designed to explain the trading institution in great detail both theoretically and using examples. To make sure they had understood the market structure, several review questions were discussed with the subjects both after they had read the first half of the instructions, dealing with the stock market, and after the second half, explaining the option market. In the next step, the z-Tree program was started and all subjects participated in two two-period test rounds to familiarize themselves with the screen layout, interface and market mechanics (app. 0.5 h).[23] Following the test rounds, one 15-period experiment was run (app. 1.5 h), after which the subjects filled in a questionnaire on their screens (app. 0.25 h), which is reprinted in Table 12 in Sect. 4.1.3. The first round was followed by a lunch break and by a second 15-period round with the same treatment design, but with starting cash and stock inventories that varied for some subjects.[24] This second round was once again followed by the subjects filling in the (same) questionnaire.[25,26] After this, subjects were asked into an extra room, where they received their payment in private. This payout consisted of the sum over all rounds R of the initial cash endowments $W_{r,t=0}$ of each round r, including a loan of € 10 (in each period), plus any proceeds from stock sales, minus any expenditures for stock purchases (subsumed in the net proceeds from stock transactions, $ST_{r,t}$), plus dividends received from shares held at the end of each period t (the dividend per share for the period, $d_{r,t}$, times the number of shares owned by the subject, $x_{r,t}$) minus all investments in options, $OI_{r,t}$, plus all proceeds from options at their maturity dates ($PO_{r=M}^{o \in \mathbb{O}_{r,t=M}}$, where $\mathbb{O}_{r,t=M}$ is the

[23] Subjects participating in the DO8 treatment could trade a digital option during period 1 of a test round, which expired at the end of period 1. In period 2, the option market was closed. Participants of the DO5/10/15 treatment could trade options in both test periods; these options expired at the end of period 1 and of period 2, respectively.

[24] See Tables 3 and 4 below for more information on subjects' initial endowments.

[25] In the first experiment, a third round was run, again followed by the questionnaire.

[26] Most similar studies play each round on a different day, which gives subjects time to recuperate and reflect on the task they are faced with (exceptions to this rule are Stanley (1997) and Haruvy et al. (2007), who play three rounds in one day). This pattern was not followed in this study to ensure that all subjects would return for the repeat rounds and that the time in which subjects could discuss the experiment would be minimized. Observations during the experiment confirmed that subjects were highly motivated to perform well in the repeat round and – having consumed lunch between the first and second rounds – did not appear to lack energy and attention. Nonetheless, it cannot be ruled out that this difference in design possibly causes differences in results.

set of options with maturity M held by the subject, where $M \in \mathbb{M}^{DO8} = \{8\}$ is the option's maturity date in the DO8 treatment and $M \in \mathbb{M}^{DO5/10/15} = \{5, 10, 15\}$ in the DO5/10/15 treatment), minus the loan of € 10 (in each round), plus a € 3 attendance fee (payable once per session). All prices in the experiment were quoted in Euro cents (€ 0.01). There were no transaction costs in either the stock or the option market. A subject's payout function from the experiment thus was (all values expressed in Euro):

$$PO = \max\left\{0, \sum_{r=1}^{R}\left[W_{r,t=0} + \sum_{t=1}^{15}(ST_{r,t} + d_{r,t} \cdot x_{r,t} - OI_{r,t} + PO_{r=M}^{o \in \mathbb{O}_{r,t=M}}) - 10\right] + 3\right\}$$

The maximum term in the payout function guaranteed that subjects could not make losses exceeding the attendance fee.

It is a well-known problem in experimental studies that the numerical values used in the instructions may significantly influence subsequent prices in the early periods. To minimize this anchoring effect,[27] the numbers used in the instructions were drawn from widely scattered areas of the value space that subjects could conceivably encounter in the experiment. A similar problem is found with regard to the numbers used in test rounds. Güth et al. (1997) for example described such an observation of anchoring with regard to the numbers they used in their training sessions. In this study, this problem was recognized and provided for as follows: For the test rounds, subjects received endowments of cash from a discrete, integer-only uniform distribution over the interval [225, 945] (in cents) and of stock from the same distribution over [1, 3] (in units). For the real rounds, each trader received one of three starting portfolios of cash and stock, each of which carried an expected value of € 13.05. These portfolios are shown in Table 3. They were chosen equal to the endowments in the futures treatment of Porter and Smith (1995, Table 2, p. 517) to ensure optimal comparability of the results. The range of test round endowments thus reflects the range of initial endowments of the real rounds, so that any anchoring tendency by subjects from the test rounds would fall into the same range of values as those the subjects would encounter

Table 3 Initial trader portfolios

Portfolio type	Initial stock (number of shares)	Initial cash (€)	Loan (€)	Expected earnings (€)
A	1	9.45	10.00	13.05
B	2	5.85	10.00	13.05
C	3	2.25	10.00	13.05

The table lists the three different initial endowments of traders. Low-cash portfolios carry a high number of shares and vice versa. Each portfolio is complemented by a € 10 loan, repayable at the end of the experiment. All portfolios have equal fundamental value

[27] Cp. Tversky and Kahneman (1974), p. 1128, for a description of the anchoring effect.

later in the real rounds. Furthermore, the random drawings of the actual endowments were chosen to make sure that different endowment levels existed between subjects and that any communication that might take place between the test rounds and the real rounds would reveal only noisy information.[28] Note that eliminating this anchoring effect completely would have required letting subjects trade in the real experiment without training sessions, a decision which could be expected to influence the results more strongly than the anchoring effect remaining after the scheme described above had been implemented.

Subjects' endowments were reinitialized after each round, following the distributions just described in case of the test rounds. The endowments were thus randomized over all subjects in the test rounds and in the first real round, and followed the mapping scheme that is reprinted in Table 4 below for the second (and third) real round. The two test rounds did not count toward the payout. The expected earnings per subject participating in a session comprising two rounds were € 29.10 (i.e., twice the expected earnings per round of € 13.50, plus the show-up fee of € 3). The period length in the test rounds varied between 5 and 6 min, the period length in the experiments between four and five. The rationale for the longer period length in the test rounds was to give subjects more time to get acquainted with the screen layout and interface. The reason for the variation of period length within individual rounds was that subjects received more time in periods with both spot and option trading than in periods where they could only trade in the spot market.

Table 4 Trader – portfolio mappings

Round:	11 Subjects		12 Subjects			13 Subjects		14 Subjects	
	1	2	1	2	3	1	2	1	2
Subject 1	C	C	C	C	B	C	C	C	C
Subject 2	C	B	C	B	A	C	B	C	B
Subject 3	C	A	C	A	C	C	A	C	A
Subject 4	A	C	C	B	B	C	B	C	B
Subject 5	A	B	A	B	A	A	C	A	C
Subject 6	A	A	A	C	C	A	B	A	B
Subject 7	B	C	A	A	B	A	A	A	A
Subject 8	B	B	A	A	A	A	C	A	C
Subject 9	B	A	B	A	C	B	C	B	C
Subject 10	B	B	B	C	B	B	B	B	B
Subject 11	B	B	B	B	A	B	A	B	A
Subject 12			B	C	C	B	A	B	A
Subject 13						B	B	B	B
Subject 14								B	B

Initial endowment for each subject, given the number of subjects, over all rounds. Portfolio types are as defined in Table 3

[28] Due to the time it took subjects to work through the instructions and to participate in the test rounds – on average about 1.5 h – a short break before the real rounds was indispensible. Naturally, it was impossible for the experimenter to prevent all communication during this break, since a number of subjects went to the toilet or purchased beverages and were thus out of his sphere of supervision for several minutes.

Chapter 4
Results

4.1 Experimental Results

The discussion of the results will start with summary statistics in Sect. 4.1.1. To put these results into perspective, the following section lists a number of definitions for bubble measures found in the literature, as well as some that were created specifically for this study. It then reports more than 600 measure results from 22 studies from the literature and from the present experiment in Tables 7 through 11 (Sect. 4.1.2.1 through 4.1.2.5). The bubble measure results for the experiment of this book are new and reported here for the first time.[1] In addition to these analytical measures, every subject answered a questionnaire at the end of each 15-period round. A translation of the questions, some summary statistics of subjects' responses, and a discussion of the most prominent findings is given in Sect. 4.1.3.

Apart from the central research question of this book – namely what effect a digital option market has on spot market efficiency – the observations made during this research project sparked the formulation of a new hypothesis regarding subjects' expectation formation mechanism in this type of asset market, as well as the discovery of some interesting behavioral patterns. These will be discussed in Sect. 4.2 below.

4.1.1 Descriptive Statistics

Table 5 lists some summary statistics on the trading activity in the experimental markets. The first content row lists the mean total number of stock transactions per round, separately for each treatment. The next three rows list the percentage

[1] Preliminary results were reported in a working paper of the author, presented at conferences in Graz, Austria, and Lille, France. However, they have not yet been reported in a publication.

S. Palan, *Bubbles and Crashes in Experimental Asset Markets*,
Lecture Notes in Economics and Mathematical Systems 626,
DOI: 10.1007/978-3-642-02147-3_4, © Springer-Verlag Berlin Heidelberg 2009

Table 5 Descriptive statistics of transactions in the experimental market

Measure	DO8				DO5/10/15			Both
Round:	1	2	3	All	1	2	All	All
Number of stock transactions	151.5	101.3	98	123.2	137	116	126.5	124.5
– below fundamental value (%)	10.9	17.5	3.1	12.6	36.5	15.8	27.0	18.5
– between fund. and max. dividend value (%)	59.6	61.7	96.9	63.7	39.7	65.5	51.5	58.7
– above maximum dividend value (%)	29.5	20.7	0.0	23.7	23.8	18.7	21.5	22.8
Number of option transactions	35.5	32.5	26	33.1	76.3	71	73.7	49.3
– below fundamental value (%)	14.8	0.0	0.0	7.0	26.2	6.1	16.5	12.7
– at fundamental value (%)	0.0	0.0	0.0	0.0	0.0	0.9	0.5	0.3
– above fundamental value (%)	85.2	100.0	100.0	93.0	73.8	93.0	83.0	87.0
Standard deviation of normalized payoffs[a]	0.84	0.60	0.51	0.71	1.23	0.76	1.02	0.85

[a]Normalization by dividing a subject's payoff by the mean payoff for the round

Statistical data on stock and option transactions, means over all sessions, by treatment and round

of all stock transactions that could be classified to have occurred at risk-averse to risk-neutral prices for the buyer (transactions below the fundamental value), at risk-loving or speculative prices for the buyer (transactions at or above the fundamental value, but at or below the maximum dividend value), and at irrational or speculative prices for the buyer (transactions above the maximum possible dividend value). Similarly, the next four rows list the total number of option transactions, as well as the percentages of transactions below, at or above the fundamental option value. The interpretation in the case of options is not as straightforward, since no clear statements can be made about subjects' motivation to enter into any particular option contract. Nonetheless, the fact that only 7 (12.7) percent of all transactions occurred at strike prices below the "rational" value – that is, the fundamental stock value at the option maturity date – in the DO8 (DO5/10/15) treatments shows that option strike prices did not fluctuate randomly around that value. Table 5 also shows that subjects preferred trading in the stock market to transactions in the option market, with a ratio of 3.72 (1.72) in the DO8 (DO5/10/15) treatment. The considerably higher number of option transactions in the DO5/10/15 treatment can be traced to the fact that in this institution, the option market was open for trading in all 15 periods, while in the DO8 treatment options could only be traded in periods 1–8. A more detailed (graphical) presentation of the stock and option transactions and their price levels compared to the fundamental value is provided in the detailed price plots in Fig. A.1 in Sect. 6.2 of the appendix.

4.1.2 Measures of Bubble Severity

Economists are people who work with numbers but who don't have the personality to be accountants.

Anonymous

A large number of measures documenting bubble extent and severity can be found in the literature. In the interest of comparability, all relevant measures were calculated for the experimental results of this study and – to put them into perspective – are complemented by bubble measure results from earlier work employing treatment designs based on Smith et al. (1988). In the case where these other studies already reported these bubble measures, they are reprinted here; in the case where they were not reported but could be calculated from the information provided in these studies (and in some cases by contacting the authors directly), they were calculated for presentation in Tables 7 through 11. The measures reported below are named by adding the initials of the authors of the papers they were first employed in to the measure designation.[2,3] In a few cases the established measures were complemented by new ones. These novel measures were created by the author and carry no initials in their designations. They are only suggested in cases where the set of measures previously employed in the literature did not address certain characteristics of bubbles (e.g., *ExtremeUnderpricing* on page 105 for prices below the minimum possible asset value), or where existing measures entailed severe deficiencies if applied to a more general set of treatments (e.g., the *DispersionRatio* on page 116). In order to present the measures in as simple a way as possible, the same symbols are used for the same underlying variables, regardless of the symbols used in the original articles. Furthermore, the presentation of the measures was homogenized by bringing similar formulas into a similar format where possible. In the case of multiple rounds played under the same treatment design, the measures reported are means of the bubble measures calculated for the individual rounds.[4] Finally, the classification scheme for the assignment of results from the literature to certain treatment

[2]In cases where modifications were made to the measures originally reported in the literature, the resulting (new) measures were designated as stemming from this dissertation (this was the case for the measures of AverageDispersion and ExtremeUnderpricing). This is to pre-empt a charge of wrongly reporting a modified measure as stemming from the original authors. Naturally, they are given due credit for the formulation of the original measures in the text.

[3]This book does not claim to provide an exhaustive list of bubble measures used in the literature, nor that results of all studies in this context are listed. Tables 7 through 11 aim to provide a comprehensive compilation of the most relevant measures employed in the past, for all studies encountered during the research for this book where such measures were reported or could be calculated. Similarly, some measures may have been employed in earlier studies than the earliest ones the author encountered, even though considerable effort was made to identify all studies relevant for this branch of the literature.

[4]This follows standard practice in the literature.

Table 6 Treatments and hypotheses

Treatment	Description	Hypothesis
1/3 experienced	1/3 of the traders are at least twice experienced, 2/3 are inexperienced	Small fraction of experienced traders prevents bubble
2/3 experienced	2/3 of the traders are at least twice experienced, 1/3 are inexperienced	Large fraction of experienced traders prevents bubble
Announcement high, preset	Uninformative announcement telling subjects that "The price is too high," chosen by experimenter prior to session	Uninformative communication influences bubble characteristics; experimenter choice prior to session conveys medium reliability
Announcement low, preset	Uninformative announcement telling subjects that "The price is too low," chosen by experimenter prior to session	Uninformative communication influences bubble characteristics; experimenter choice prior to session conveys medium reliability
Announcement high, random	Uninformative announcement telling subjects that "The price is too high"; randomly chosen prior to session	Uninformative communication influences bubble characteristics; random choice prior to session conveys low reliability
Announcement low, random	Uninformative announcement telling subjects that "The price is too low"; randomly chosen prior to session	Uninformative communication influences bubble characteristics; random choice prior to session conveys low reliability
Announcement true	Uninformative announcement telling subjects that "The price is too high/low"; announcement is always true, conditional on previous period's price	Informative communication influences bubble characteristics; conditional choice based on actual prices conveys high reliability
Asset rich	Ratio of the difference of total initial share value minus total cash endowments, divided by the total share value is between 0.125 and 0.5.	Less cash in the experiment deflates transaction prices
Baseline	Declining dividend value	Rational expectations equilibrium causes trading at fundamental values
Brokerage fees	Buyer and seller in a transaction pay 10 cents each for a trade	Fewer transactions due to cost of trading
Call auction	Call auction instead of double auction	Less public information decreases bubbles by decreasing speculation
Cash rich	Ratio of the difference of total initial share value minus total cash endowments, divided by the total share value is between -1 and -0.8125	More cash in the experiment inflates transaction prices
Constant value	Security pays a dividend with a mean of zero at the end of each period	Constant fundamental value decreases bubbles due to simplified convergence process

(continued)

Table 6 (continued)

Treatment	Description	Hypothesis
Dividend certainty	Security pays a fixed and known dividend amount	Trading based on dividend risk preference is eliminated
Dividend deferred	Subjects are entitled to a dividend, but payout is deferred until the end of the experimental round	Deferred dividend payment reduces liquidity, thereby deflating transaction prices
Dividend heterogeneity	Dividend level different across investors	Heterogeneous dividends increase propensity to trade and permit measurement of allocational efficiency
Dividend mix	Security pays dividends at the end of each period and an additional dividend at the end of the trading horizon	Dividend concentration focuses attention on longer-term income stream
Dividend once	Security pays a single dividend at the end of the trading horizon	Dividend concentration focuses attention on longer-term income stream
Dividend spread high	Security pays a period-end dividend of {0,8,28,98} with equal probability	Higher dividend variability compared to *Baseline* treatment increases bubble extent
Equal endowments	Homogeneous initial amounts of cash and shares over all traders	Traders do not need to balance portfolios
Experienced business	Half of all traders are twice experienced business majors, half are inexperienced arts and sciences students	Business and economics education improves market efficiency
Experienced non-business	Half of all traders are inexperienced business majors, half are twice experienced arts and sciences students	Business and economics education improves market efficiency
Futures	Agents can trade a mid-horizon (period 8) security in advance	Futures contracts should hasten the formation of common expectations
Increasing value	Security has a terminal value and pays a negative expected dividend at the end of each period	Increasing fundamental value leads to underpricing
Information on each period	Traders can buy information about dividend values for every period at the beginning of each (of three) five-period sequence	Private information reinforces common expectations and weakens reliance on information in prices
Information on periods 1, 6, and 11	Traders can buy information about dividend values at the beginning of each (of three) five-period sequences	Private information reinforces common expectations and weakens reliance on information in prices
Informed insiders	Informed traders have read Smith et al. (1988) and are given information on bids, offers and excess bids	Informed traders aware of bubble characteristics eliminate bubble
Limit price change rule	Asset price can only change by a limited amount from the previous period closing price	Suppressed expectation of rapid price changes reduces price volatility

(continued)

Table 6 (continued)

Treatment	Description	Hypothesis
Short horizon	Dividend payment date lies beyond subjects' investment horizon	Indeterminate price paths due to difficulty of backward induction
Lottery asset	A market for an asset with a high dividend with low probability accompanies the standard market	Lottery assets increase the extent and frequency of bubbles
Margin buying	Traders are given an interest-free loan to be paid back by the last period	Purchases can be leveraged to raise prices that are below dividend value
Non-business	Traders are freshman arts and sciences students	Business and economics education improves market efficiency
No speculation	Traders are either buyers or sellers, but never both	Impossibility of reaping capital gains prevents speculative bubble
Open book	All orders are visible to all participants	Information diminishes bubble size
Peak	Traders experience a time of rising, followed by a time of falling fundamental values	Bubbles are path-dependent
Rekindle	Traders twice experienced in *Baseline* design are faced with *Cash rich & Dividend spread high* parameterization	Shock in market parameters rekindles bubble even with experienced subjects
Reverse futures	Futures markets for each period, opening in reverse order of maturity	Opening futures markets from future backwards aids subjects' backward induction
Short selling $\leq$ n units	Traders may be short up to n shares of the experimental asset	Traders can leverage sales and counter price run-ups
Short selling $\leq$ n units, cash times m	Traders may be short up to n shares of the experimental asset, initial cash balance m times as great	Traders can leverage sales and counter price run-ups
Short selling $\leq$ n units, no dividends	Traders may be short up to n shares of the experimental asset, no dividend payments due on shorted shares	Traders can leverage sales and counter price run-ups
Short selling x% cash reserve	Traders may be short up to a value equivalent to x% of their cash reserve	Traders can leverage sales and counter price run-ups
Short selling x% cash reserve, cash times m	Traders may be short up to a value equivalent to x% of their cash reserve, initial cash balance m times as great	Traders can leverage sales and counter price run-ups
Short selling flexible cash reserve	Traders may be short up to a value equivalent to 100% (200%) of their cash reserve if last transaction price was above (below) fundamental value	Traders can leverage sales and counter price run-ups

(continued)

Table 6 (continued)

Treatment	Description	Hypothesis
Switch	Traders trained in two sessions of dividend certainty, then subjected to third session of baseline design	Dividend uncertainty causes bubble with subjects twice experienced in dividend certainty treatment
Symmetric dividend	Five-point discrete dividend distribution with symmetric probabilities	Symmetric dividend focuses attention on expected value
Two markets	Trade in a service market in addition to asset market	Second trading outlet reduces price inefficiencies due to Active Participation Hypothesis (the APH is defined in Lei et al. 2001)
Valley	Traders experience a time of falling, followed by a time of rising fundamental values	Bubbles are path-dependent

This table lists the general treatment designs employed in studies on the performance of experimental asset markets in the literature. The treatments are listed with a short designation, with a brief description of the experimental institution employed, and with the hypothesis the treatment was designed to test

classes is based on the scheme employed in Porter and Smith (1994).[5] Table 6 gives an overview of these general groups of treatments which are used to structure the presentation of the bubble measure results in the following sections.[6]

Apart from giving the reader the chance to compare the bubble measure results for this work to outcomes from earlier studies, this structured presentation of the information contained in a total of more than 600 measure results from 22 experimental studies may help future researchers to discuss their observations before the background of these findings.[7]

[5]Cp. Porter and Smith (1994), Table 1, p. 116.

[6]In experimental studies, no two experiments follow exactly the same institution. For this reason, results designated as stemming from e.g., a *Baseline* treatment might have been run with a different subject base, in a different language, might employ slightly different dividend distributions, have different instructions and be conducted on a different software platform from previous studies. Since such differences are inevitable when viewing experimental work from a high-level perspective, the reported statistics intentionally sacrifice detail to gain homogeneity of presentation.

[7]An observation that supports this hope is the example of Haruvy and Noussair (2006), who in their Table 4 on p. 1135, reported – among others – an amplitude value of 1.53 for the Porter and Smith (1995) baseline treatments, which is the same number as given in the original article of Porter and Smith. However, Porter and Smith employed the amplitude definition designated AmplitudeK below, while Haruvy and Noussair defined amplitude as in AmplitudeHN below, which makes the comparison of the reported measures hard to interpret. Hopefully, the bubble measure presentation provided here will simplify the comparison of measure results across different studies in the future.

4.1.2.1 Amplitude Measures

Amplitude measures generally report the difference between the lowest and highest mean period price in an experimental round, which may then be normalized using some form of fundamental value. Haruvy and Noussair (2006) employed the following measure:

$$\text{AmplitudeHN} = \max_t\left(\frac{\overline{P}_t - f_t}{f_t}\right) - \min_t\left(\frac{\overline{P}_t - f_t}{f_t}\right) \tag{4.2}$$

where $\overline{P}_t$ is the mean transaction price in the stock market in period t, and f_t is the fundamental or dividend holding value in the same period. In a very similar approach, King (1991) used the following measure:

$$\text{AmplitudeK} = \max_t\left(\frac{\overline{P}_t - f_t}{f_1}\right) - \min_t\left(\frac{\overline{P}_t - f_t}{f_1}\right) \tag{4.3}$$

The value f_1 is the asset's fundamental value in the first period, which in the experiment reported in this book equaled 360. As a final amplitude measure, Van Boening et al. (1993) reported values calculated without normalization using a proxy for the fundamental value:

$$\text{AmplitudeVWL} = \max_t(\overline{P}_t - f_t) - \min_t(\overline{P}_t - f_t) \tag{4.4}$$

Clearly, results of formula (4.4) can easily be transformed into results of the form of formula (4.3). Since formula (4.3) has become the norm for papers later than 1993 (except for those reporting AmplitudeHN), all results calculated using formula (4.4) are reported as transformed into the formula (4.3) version.[8]

Table 7 below reports the results of bubble amplitude measurements taken from a number of different studies. The comparison of the AmplitudeHN results for inexperienced subjects from the literature with those derived for this work shows that the bubble in this study's experiment was relatively large, with only 1 out of 18 results from the literature exceeding the 5.676 (6.972) calculated for the DO8 (DO5/10/15) treatments. The verdict on the AmplitudeK measure results is similar, yet in combination with the earlier findings reveals some interesting patterns. First off, experimental markets employing call auction institutions seem to generally produce high-amplitude bubbles. Another institutional feature that seems to be associated with high amplitudes is margin buying, or – more generally – large

[8] This transformation is possible, because all studies in the literature report the fundamental value in the first period. Such a transformation is not possible between formulas (2) and (3) or (2) and (4), because – while the fundamental value process is usually reported in studies in the literature – the period where the deviation of the price from the fundamental value reached its maximum and minimum is not usually reported.

Table 7 Amplitude measures

Measure Article / Treatment	Experience None	 Once	 Twice	 n
AmplitudeHN (4.2) – Haruvy and Noussair (2006)				
Caginalp et al. (1998) – Asset rich & Dividend once	0.380	–	–	4;0;0
— Cash rich & Dividend once	0.923	–	–	3;0;0
Caginalp et al. (2001) – Asset rich & Call auction	2.664	–	–	5;0;0
— Asset rich & Call auction & Constant value	0.468	–	–	5;0;0
— Asset rich & Call auction & Dividend deferred	3.275	–	–	4;0;0
— Asset rich & Call auction & Dividend deferred & Open book	1.386	–	–	3;0;0
— Asset rich & Call auction & Open book	3.945	–	–	4;0;0
— Call auction	2.354	–	–	2;0;0
— Call auction & Cash rich	10.449	–	–	3;0;0
— Call auction & Cash rich & Constant value	0.334	–	–	6;0;0
— Call auction & Cash rich & Open book	8.290	–	–	4;0;0
— Call auction & Constant value	0.333	–	–	1;0;0
Davies (2006) – Baseline	3.007	–	–	3;0;0
— Increasing value	0.873	–	–	7;0;0
Haruvy et al. (2007) – Call auction	8.83	2.87	1.82	6;6;6
Haruvy and Noussair (2006) – Baseline	2.61	–	–	2;0;0
— Short selling ≤ 3 units	1.00	–	–	2;0;0
— Short selling ≤ 6 units	1.55[a]	0.69	–	3;3;0
— Short selling ≤ 6 units, cash times 10	5.73	–	–	2;0;0
— Short selling 100% cash reserve	1.46[a]	1.06	–	3;3;0
— Short selling 100% cash reserve, cash times 10	5.12	–	–	2;0;0
— Short selling 150% cash reserve	1.29	–	–	2;0;0
— Short selling flexible cash reserve	1.18	–	–	2;0;0
Hirota and Sunder (2007) – Dividend once & -certainty	1.000	–	–	1;0;0
— Dividend once & -certainty & -heterogeneity	0.949	–	–	4;0;0
— Dividend once & -certainty & -heterogeneity & short horizon	3.944	–	–	4;0;0
— Dividend once & short horizon	4.653	–	–	2;0;0
Noussair and Powell (2008) – Peak	4.372	6.314	5.088	5;5;5
— Valley	7.372	5.262	4.400	5;5;5
Noussair et al. (2001) – Constant value	0.515	–	–	8;0;0
This study - DO8	*5.676*	*1.917*	*0.800*	*4;4;1*
—DO5/10/15	*6.972*	*2.392*	–	*3;3;0*
AmplitudeK (4.3) – King (1991)				
Ackert and Church (2001) – Baseline	1.07	0.52	–	3;2;0
— Experienced business	0.56	–	–	2;0;0
— Experienced non-business	0.86	–	–	2;0;0
— Non-business	1.21	0.67	–	3;2;0
Caginalp et al. (1998) – Asset rich & Dividend once	0.380	–	–	4;0;0
— Cash rich & Dividend once	0.923	–	–	3;0;0
Caginalp et al. (2001) – Asset rich & Call auction	1.056	–	–	5;0;0
— Asset rich & Call auction & Constant value	0.005	–	–	5;0;0
— Asset rich & Call auction & Dividend deferred	1.241	–	–	4;0;0
— Asset rich & Call auction & Dividend deferred & Open book	0.868	–	–	3;0;0

(continued)

Table 7 (continued)

Measure Article / Treatment	Experience			
	None	Once	Twice	n
— Asset rich & Call auction & Open book	1.347	–	–	4;0;0
— Call auction	0.876	–	–	2;0;0
— Call auction & Cash rich	2.598	–	–	3;0;0
— Call auction & Cash rich & Constant value	0.003	–	–	6;0;0
— Call auction & Cash rich & Open book	1.712	–	–	4;0;0
— Call auction & Constant value	0.003	–	–	1;0;0
Corgnet et al. (2008) – Announcement high, preset	0.987	0.843	–	3;3;0
— Announcement low, preset	1.256	1.208	–	3;3;0
— Announcement high, random	1.17	–	–	3;3;0
— Announcement low, random	1.14	–	–	3;3;0
— Announcement true	1.07	–	–	3;3;0
— Baseline	1.095	0.930	–	3;3;0
Davies (2006) – Baseline	1.010	–	–	3;0;0
— Increasing value	1.530	–	–	7;0;0
Dufwenberg et al. (2005) – Baseline	0.815	0.798	0.593	10;10;10
— 1/3 experienced	0.580	–	–	5;0;0
— 2/3 experienced	0.515	–	–	5;0;0
Hirota and Sunder (2007) – Dividend once & -certainty	1.000	–	–	1;0;0
— Dividend once & -certainty & -heterogeneity	0.949	–	–	4;0;0
— Dividend once & -certainty & -heterogeneity & short horizon	3.944	–	–	4;0;0
— Dividend once & short horizon	4.653	–	–	2;0;0
Hussam et al. (2008) – Baseline	1.237[b]	0.808[b]	0.229[b]	34;13;8
— Cash rich & Dividend spread high	1.473[b]	1.316[b]	1.030[b]	3;3;3
— Rekindle	1.142[b]	–	–	3;0;0
King (1991) – Information on each period	0.726	–	–	6;0;0
— Information on periods 1, 6, and 11	0.696	–	–	6;0;0
King et al. (1993) – Baseline	0.344	0.213	0.030	10;3;2
— Brokerage fees	0.203	0.175	–	2;3;0
— Equal endowments	0.519	–	–	4;0;0
— Informed insiders	0.174	0.101	–	1;1;0
— Informed insiders & short selling	0.264	0.071	–	1;2;0
— Limit price change rule	1.042	0.492	0.193	2;2;2
— Margin buying	1.011	0.319	–	1;1;0
— Short selling $\leq$ 2 units, no dividends	0.447	0.214	0.110	4;5;3
— Short selling $\leq$ 2 units, no dividends & margin buying	0.243	0.179	–	1;1;0
Noussair et al. (2001) – Constant value	0.515	–	–	8;0;0
Noussair and Tucker (2006) – Reverse futures	0.331	–	–	4;0;0
Porter and Smith (1994) – Baseline	1.21	0.75	0.10	19;4;3
— Brokerage fees	0.73	0.63	–	2;3;0
— Dividend certainty	1.10	0.52	–	3;3;0
— Equal endowments	1.87	–	–	4;0;0
— Futures	0.92	0.60	–	3;2;0
— Informed insiders	0.63	0.25	–	2;3;0
— Limit price change rule	2.51	1.77	0.70	2;2;2
— Margin buying	3.64	1.15	–	2;1;0
— Short selling $\leq$ 2 units	1.61	0.76	0.40	4;5;3

(continued)

Table 7 (continued)

Measure Article / Treatment	Experience			
	None	Once	Twice	n
Porter and Smith (1995) – Baseline	1.53	0.86	–	10;8;0
— Dividend certainty	1.09	0.52	–	3;3;0
— Futures & margin buying	0.92	0.60	–	3;2;0
— Margin buying	3.21	1.12	–	3;1;0
— Switch	–	–	0.40	0;0;2
Smith et al. (1988) – Baseline	1.24			
Smith et al. (2000) – Baseline	1.388	0.927	–	6;3;0
— Dividend mix	0.925	0.428	–	5;2;0
— Dividend once	0.684	0.305	–	8;2;0
Van Boening et al. (1993) – Call auction	1.609	1.183	0.481	2;2;2
— Symmetric dividend	0.627	0.701	0.253	2;2;2
This study - DO8	*1.238*	*0.933*	*0.461*	*4;4;1*
—DO5/10/15	*1.572*	*1.016*	–	*3;3;0*

[a]Calculated by combining Tables 2 and 3 of Haruvy and Noussair (2006)

[b]Results stem from a seemingly unrelated regression (SUR), not from averaging this measure directly over all applicable rounds

This table compares amplitude measure results reported in the literature for experimental asset markets. Results for the DO8 and DO5/10/15 treatments are printed in italics at the bottom of each block. The bubble measure names are followed by their formula number and the article they were first proposed in. All numbers reported in columns 2–4 are means over all rounds conducted with the stated treatment and level of experience. The last column lists the number of rounds the reported measures are the mean of, for each level of experience. When calculated specifically for this table, measures are listed with three digits of precision

liquidity, as in the *Cash rich* treatments. This might be one reason for the relatively high bubble measure results for the present study, since the subjects were endowed with a margin buying capacity (loan) of € 10 in each round. Overall, the amplitude results provide no support for the research hypothesis that a digital option market would improve efficiency in the experimental stock market. Furthermore, they indicate that the DO5/10/15 treatment produced price paths that were even less efficient than the DO8 treatment, a pattern that also holds (though not strictly) for the other bubble measures reported below.

4.1.2.2 Deviation Measures

Deviation measures are related to amplitude measures, in that both calculate a metric for the discrepancy between the observed transaction prices and the underlying fundamental value. While amplitude measures focus on the difference between maximum and minimum price (deviation), the measures presented in this section report on the deviation of prices from the fundamental or from a theoretical maximum or minimum value of the experimental asset. An example of the former (i.e., a measure of deviation from the fundamental value) was introduced in King et al. (1993) and Van Boening et al. (1993), who calculated a measure of normalized absolute price deviation which summed the deviation of the stock price from the fundamental value for every transaction in every period, and normalized it by dividing by the total number of shares outstanding:

$$\text{DeviationKSWV} = \frac{\sum_{t=1}^{T} \sum_{i_t=1}^{I_t} |P_{i_t} - f_t|}{q} \tag{4.5}$$

In this formula, P_{i_t} is the transaction price of transaction i in period t, where I_t is the total number of transactions in period t, and q is the total number of shares outstanding in the experimental round, sometimes also referred to as the *total stock of units*, TSU (i.e., $2 \cdot n$ in this book's experiment, where n is the number of subjects, since each subject held an average of two shares of stock). In this calculation, they used prices quoted in dollars.[9]

A similar but distinct measure is the Total Dispersion, introduced by Haruvy and Noussair (2006). It is calculated as follows:

$$\text{TotalDispersion} = \sum_{t=1}^{T} |\tilde{P}_t - f_t|$$

Total Dispersion gives an indication of the absolute deviation of median prices ($\tilde{P}_t$) from fundamental values. The same article also proposes a related measure, the Average Bias:

$$\text{AverageBiasHN} = \frac{\sum_{t=1}^{T} (\tilde{P}_t - f_t)}{T} \tag{4.6}$$

AverageBiasHN gives an indication of the average per-period deviation of prices from fundamental values. In other words, it contains information about the average distance between the mean period price and the fundamental value, with a value of zero indicating that average period prices perfectly tracked the fundamental value, while values larger (smaller) than zero indicate positive (negative) departures from the fundamental value process, implying bubbles (negative bubbles).[10] Since the normalization over the number of periods per round makes this measure more versatile than the Total Dispersion, the latter will be reported in a similar form in this dissertation:

$$\text{AverageDispersion} = \frac{\sum_{t=1}^{T} |\tilde{P}_t - f_t|}{T} \tag{4.7}$$

[9]Later studies, e.g. Haruvy and Noussair (2006), mostly quoted prices in cents. They then calculated an equivalent measure as follows: $\left(\sum_{t=1}^{T} \sum_i |P_{i,t} - f_t|\right)/(100 \cdot q)$.

[10]Corgnet et al. (2008) proposed a "Bias" measure, which is calculated as $\sum_{t=1}^{T} |\tilde{P}_t - f_t|/T$, which is a simple transformation of TotalDispersionHN and thus contains equivalent information to the latter.

Lei et al. (2001) and Ackert et al. (2006a) reported information on the number of transactions conducted at prices above (below) the maximum (minimum) possible remaining dividend payoff of the stock over the remaining time in the experimental round for their experiments. This information can be captured by measures defined as follows:[11]

$$\text{OverpricedTransactions} = \frac{\sum_{t=1}^{T} \sum_{i_t=1}^{I_t} x_{i_t}^{\max}}{q^r} \tag{4.8}$$

where $x_{i_t}^{\max} = \begin{cases} 0 & \text{if } P_{i_t} \leq f_t^{\max} \\ 1 & \text{if } P_{i_t} > f_t^{\max} \end{cases}$

and

$$\text{UnderpricedTransactions} = \frac{\sum_{t=1}^{T} \sum_{i_t=1}^{I_t} x_{i_t}^{\min}}{q^r} \tag{4.9}$$

where $x_{i_t}^{\min} = \begin{cases} 0 & \text{if } P_{i_t} \geq f_t^{\min} \\ 1 & \text{if } P_{i_t} < f_t^{\min} \end{cases}$

For both, q^r is the total number of transactions in the experimental round, $f_t^{\max}$ ($f_t^{\min}$) is the maximum (minimum) possible remaining dividend payoff from one share from period t until the end of the round, and $x_{i_t}^{\max}$ ($x_{i_t}^{\min}$) is a binary variable indicating whether transaction i in period t was conducted at a price above $f_t^{\max}$ (below $f_t^{\min}$). Ackert and Church (2001) defined a rather similar measure of the number of periods in which the mean period price exceeded the maximum remaining dividend payoff from one share from period t until the end of the session as follows:

$$\text{ExtremeOverpricingAC} = \sum_{t=1}^{T} x_t^{\text{over}} \tag{4.10}$$

where $x_t^{\text{over}} = \begin{cases} 0 & \text{if } \overline{P}_t \leq f_t^{\max} \\ 1 & \text{if } \overline{P}_t > f_t^{\max} \end{cases}$

An analogous measure for underpricing would be:

$$\text{ExtremeUnderpricing} = \sum_{t=1}^{T} x_t^{\text{under}} \tag{4.11}$$

where $x_t^{\text{under}} = \begin{cases} 0 & \text{if } \overline{P}_t \geq f_t^{\min} \\ 1 & \text{if } \overline{P}_t < f_t^{\min} \end{cases}$

[11] Note that the measure "OverpricedTransactions" tests for the presence of a *strong bubble*, as introduced in Sect. 2.1.3 and defined in Allen et al. (1993).

Table 8 Deviation measures

Measure Article / Treatment	Experience None	Once	Twice	n
AverageBiasHN (4.6) – Haruvy and Noussair (2006)				
Noussair and Powell (2008) – Peak	76.902	52.282	31.166	5;5;5
— Valley	67.480	88.146	75.808	5;5;5
This study - DO8	*158.300*	*141.833*	*75.567*	*4;4;1*
—DO5/10/15	*133.156*	*157.044*	–	*3;3;0*
AverageDispersion (4.7) – This dissertation				
Corgnet et al. (2008) – Announcement high, preset	4.31[a]	–	–	3;3;0
— Announcement low, preset	6.60[a]	–	–	3;3;0
— Announcement high, random	10.70[a]	–	–	3;3;0
— Announcement low, random	9.56[a]	–	–	3;3;0
— Announcement true	8.44[a]	–	–	3;3;0
— Baseline	11.16[†]	–	–	3;3;0
Noussair and Powell (2008) – Peak	83.540	56.787	37.593	5;5;5
— Valley	96.813	93.373	80.700	5;5;5
This study - DO8	*171.967*	*144.700*	*75.567*	*4;4;1*
—DO5/10/15	*178.533*	*165.244*	–	*3;3;0*
DeviationKSWV (4.5) – King et al. and Van Boening et al. (1993)				
Corgnet et al. (2008) – Announcement high, preset	0.585	0.827	–	3;3;0
— Announcement low, preset	1.288	1.210	–	3;3;0
— Announcement high, random	1.16	–	–	3;3;0
— Announcement low, random	1.13	–	–	3;3;0
— Announcement true	1.054	–	–	3;3;0
— Baseline	1.301	0.843	–	3;3;0
Dufwenberg et al. (2005) – Baseline	1.667	1.612	0.813	10;10;10
— 1/3 experienced	1.261	–	–	5;0;0
— 2/3 experienced	0.866	–	–	5;0;0
Haruvy et al. (2007) – Call auction	2.19	1.29	0.53	6;6;6
King et al. (1993) – Baseline	5.68	2.77	0.279	10;3;2
— Brokerage fees	3.91	1.51	–	2;3;0
— Equal endowments	13.57	–	–	4;0;0
— Informed insiders	1.61	0.691	–	1;1;0
— Informed insiders & short selling	3.05	1.21	–	1;2;0
— Limit price change rule	9.46	2.12	0.390	2;2;2
— Margin buying	15.30	2.61	–	1;1;0
— Short selling $\leq$ 2 units	11.88	3.90	1.23	4;5;3
— Short selling $\leq$ 2 units & margin buying	16.30	9.71	–	1;1;0
Noussair and Powell (2008) – Peak	4.658	2.806	1.130	5;5;5
— Valley	6.116	4.488	3.362	5;5;5
Noussair et al. (2001) – Constant value	2.24	–	–	8;0;0
Noussair and Tucker (2006) – Reverse futures	0.239	–	–	4;0;0
Smith et al. (2000) – Baseline	5.498	0.182	–	6;3;0
— Dividend mix	3.596	-1.150	–	5;2;0
— Dividend once	1.861	-1.681	–	8;2;0
Van Boening et al. (1993) – Call auction	5.083	1.100	0.365	2;2;2

(continued)

Table 8 (continued)

Measure Article / Treatment	Experience			
	None	Once	Twice	n
— Symmetric dividend	5.385	4.125	1.130	2;2;2
This study - DO8	*11.483*	*6.934*	*3.581*	*4;4;1*
— DO5/10/15	*9.140*	*7.289*	–	*3;3;0*
ExtremeOverpricingAC (4.10) – Ackert and Church (2001)				
Ackert and Church (2001) – Baseline	1.667	0.000	–	3;2;0
— Experienced business	0.000	–	–	2;0;0
— Experienced non-business	0.000	–	–	2;0;0
— Non-business	0.333	0.000	–	3;2;0
Caginalp et al. (1998) – Asset rich & Dividend once	0.250	–	–	4;0;0
— Cash rich & Dividend once	1.333	–	–	3;0;0
Caginalp et al. (2001) – Asset rich & Call auction	0.800	–	–	5;0;0
— Asset rich & Call auction & Constant value	0.000	–	–	5;0;0
— Asset rich & Call auction & Dividend deferred	1.500	–	–	4;0;0
— Asset rich & Call auction & Dividend deferred & Open book	0.000	–	–	3;0;0
— Asset rich & Call auction & Open book	3.500	–	–	4;0;0
— Call auction	1.000	–	–	2;0;0
— Call auction & Cash rich	7.333	–	–	3;0;0
— Call auction & Cash rich & Constant value	0.000	–	–	6;0;0
— Call auction & Cash rich & Open book	6.500	–	–	4;0;0
— Call auction & Constant value	0.000	–	–	1;0;0
Davies (2006) – Baseline	4.000	–	–	3;0;0
— Increasing value	0.000	–	–	7;0;0
Hirota and Sunder (2007) – Dividend once & -certainty	0.000	–	–	1;0;0
— Dividend once & -certainty & -heterogeneity	2.500	–	–	4;0;0
— Dividend once & -certainty & -heterogeneity & short horizon	1.132	–	–	4;0;0
— Dividend once & short horizon	7.500	–	–	2;0;0
This study - DO8	*4.500*	2.250	0.000	*4;4;1*
— DO5/10/15	*5.000*	*3.333*	–	*3;3;0*
ExtremeUnderpricing (4.11) – This study				
Caginalp et al. (1998) – Asset rich & Dividend once	2.500	–	–	4;0;0
— Cash rich & Dividend once	0.000	–	–	3;0;0
Caginalp et al. (2001) – Asset rich & Call auction & Constant value	5.800	–	–	5;0;0
— Call auction & Cash rich & Constant value	0.333	–	–	6;0;0
— Call auction & Constant value	0.000	–	–	1;0;0
Davies (2006) – Baseline	0.000	–	–	3;0;0
— Increasing value	6.571	–	–	7;0;0
Hirota and Sunder (2007) – Dividend once & -certainty	8.000			1;0;0

(continued)

Table 8 (continued)

Measure Article / Treatment	Experience None	 Once	 Twice	 n
— Dividend once & -certainty & -heterogeneity	1.250			4;0;0
— Dividend once & -certainty & -heterogeneity & short horizon	2.000			4;0;0
— Dividend once & short horizon	7.500			2;0;0
This study - DO8	*n/a*	*n/a*	*n/a*	*n/a*
— DO5/10/15	*n/a*	*n/a*	–	*n/a*
OverpricedTransactions (4.8) – This study				
Ackert et al. (2006a) – Baseline & lottery asset	0.307	–	–	5;0;0
— Margin buying & lottery asset	0.435	–	–	4;0;0
— Short selling $\leq$ 5 units & lottery asset	0.179	–	–	4;0;0
Lei et al. (2001) – No speculation	0.376	–	–	3;0;0
— Two markets & No speculation	0.462	–	–	3;0;0
This study - DO8	*0.279*	*0.186*	0.000	*4;4;1*
— DO5/10/15	*0.244*	*0.232*	–	*3;3;0*
UnderpricedTransactions (4.9) – This study				
Ackert et al. (2006a) – Baseline & lottery asset	0.043	–	–	5;0;0
— Margin buying & lottery asset	0.045	–	–	4;0;0
— Short selling $\leq$ 5 units & lottery asset	0.268	–	–	4;0;0
Lei et al. (2001) – No speculation	0.161	–	–	3;0;0
— Two markets & No speculation	0.010	–	–	3;0;0
This study – DO8	*n/a*	*n/a*	*n/a*	*n/a*
— DO5/10/15	*n/a*	*n/a*	–	*n/a*

[a]Corgnet et al. (2008) use the mean period transaction price instead of the median in their calculation of AverageDispersion

This table compares deviation measure results reported in the literature for experimental asset markets. Results for the DO8 and DO5/10/15 treatments are printed in italics at the bottom of each block. The bubble measure names are followed by their formula number and the article they were first proposed in. All numbers reported in columns 2–4 are means over all rounds conducted with the stated treatment and level of experience. The last column lists the number of rounds the reported measures are the mean of, for each level of experience. When calculated specifically for this table, measures are listed with three digits of precision

The difference between the measures (4.8) and (4.10), as well as that between (4.9) and (4.11) is that (4.8) and (4.9) are based on prices outside the possible ex post fundamental value range for individual transactions, while (4.10) and (4.11) consider them at the mean period price level.[12]

[12]An additional difference is that the Ackert et al. (2006) measure is not normalized over the number of periods in a session, a modification that would facilitate comparison across experiments with different numbers of periods. However, transforming the extreme over- and underpricing measures reported in Table 8 to reflect such a modification can easily be accomplished. All it requires is dividing the reported measure results by 15, the number of periods in the Ackert et al. (2006) and the DO8 and DO5/10/15 experiments.

Table 8 lists the deviation measure results for all articles they could be obtained or calculated for. The findings for the first measure, DeviationKSWV, resemble those from the previous section on bubble amplitude. Once again, treatments giving subjects the opportunity to buy stock on margin exhibit large price inefficiencies, a fate that the experiment reported here is not spared. While this effect diminishes for the once experienced subjects in all studies, this reduction is less pronounced for the present experiment than for those from the literature. However, due to the higher variance of the measure results both overall and within individual treatment categories (*Baseline*, *Call auction*), the robustness of the results must be considered less certain than before.

Moving on to the measure of ExtremeOverpricingAC, the evidence from comparing this study with the results from the literature becomes mixed. While some treatments exhibit a lower mean number of periods in which the mean transaction price exceeds the maximum asset value, some also exhibit a higher frequency of such periods. Nonetheless, the simple fact that mean period prices exceeded the maximum dividend value of the stock in 4.5 (5) out of 15 periods in the DO8 (DO5/10/15) treatment is solid evidence of an impressive bubble in these markets. The measure taking the opposite view – ExtremeUnderpricing – is reported for other studies, but not for this study. The reason is that in the latter, the minimum possible value of a share of stock was zero and transaction prices in the experiment were constrained to be strictly positive, thus precluding transactions below the minimum fundamental value.

The remaining two measures discussed above, OverpricedTransactions and UnderpricedTransactions, look at trades at prices outside the bounds of possible ex post fundamental values at the level of individual transactions. In the setting of Lei et al. (2001), these measures provided clear evidence of irrationality, because capital gains were impossible. On the contrary, subjects in the DO8 and DO5/10/15 designs were free to purchase and sell stock at any time during the experiment, which gave them the possibility to reap capital gains. However, Harrison and Kreps (1978) asserted that – even when capital gains *are* possible – an investor can be referred to as exhibiting speculative behavior if she is willing to buy an asset she would not be willing to buy if she were obliged to hold it forever.[13] In other words, a speculator is willing to pay more for an asset which she can resell than for an asset she is required to hold forever. This lends relevance to the OverpricedTransaction and UnderpricedTransaction measures even in settings with the possibility of capital gains. In contrast to the negative news from the previous sections, the results regarding the OverpricedTransaction measure are relatively encouraging.[14] The percentage of transactions at prices above the maximum dividend value – with values of 27.9% (24.4%) in the DO8 (DO5/10/15) treatments – was lower than four out of five results from the literature. Nonetheless, it might still be interesting to note that the results reported in Table 8 for the DO8 and DO5/10/15 experiments show that the mean over both treatments of the results of the OverpricedTransaction

[13] Cp. Harrison and Kreps (1978), p. 323.

[14] The measure of UnderpricedTransactions once again does not apply to the setting of the experiment presented here.

measure in rounds with inexperienced (once experienced) subjects was an impressive 26.15% (20.90%) – figures which cannot be reconciled with a market that is even close to being informationally efficient.

4.1.2.3 Duration Measures

King (1991) reported a boom duration measure that is calculated as the "Number of periods from low to high mean price," where the mean price was defined as "Mean contract price (measured as difference from expected price)...".[15] Strictly speaking, the definition is ambiguous in three respects: On the one hand, it could mean either that first the high mean price period is located and then the time from the low mean price period *prior* to the high mean price period is calculated (That is: $[k: \overline{P}_k - f_t = \max_t(\overline{P}_t - f_t)] - [l: \overline{P}_l - f_t = \min_{t<k}(\overline{P}_t - f_t)]$), or that the measure is simply the absolute time difference between the high and low mean price periods, regardless of which is first. (That is: $|[k: \overline{P}_k - f_t = \max_t(\overline{P}_t - f_t)] - [l: \overline{P}_l - f_t = \min_t(\overline{P}_t - f_t)]|$.) Contacted to clarify this ambiguity, Ron King stated his belief that he had used the first option. Consequently, this first specification was employed here. On the other hand, the definition is ambiguous in the case of more than one period of (equal) maximum or (equal) minimum mean price. To obtain unambiguous results, the minimum and maximum mean price periods yielding the maximum DurationK measures were chosen for the calculations reported in Table 9. Finally, King did not state whether he conditioned on the difference between mean period price and fundamental value being positive (implying overvaluation). Consequently, such a conditioning was also not conducted here. Together, these specifications imply the following formula:

$$\text{DurationK} = \max_k[k: \overline{P}_k - f_t = \max_t(\overline{P}_t - f_t)] - [l: \overline{P}_l - f_t = \min_{t<k}(\overline{P}_t - f_t)] \quad (4.12)$$

In a different approach, Porter and Smith (1995) calculated the following measure for the temporal length of the stock price bubble:

$$\text{DurationPS} = \max_{t,m}(m: \overline{P}_t - f_t < \overline{P}_{t+1} - f_{t+1} < \cdots < \overline{P}_{t+m} - f_{t+m}) \quad (4.13)$$

This formula defines the duration of a bubble as the longest uninterrupted interval during which the deviation of mean period prices from period fundamental values increased.

In their Positive duration measure, Ackert et al. (2006a) modified DurationPS by regarding only those periods where the increase in the difference between price and fundamental value produced a price in the next period that exceeded the fundamental value:

[15] King (1991), p. 203.

Table 9 Duration measures

Measure Article / Treatment	Experience			
	None	Once	Twice	n
DurationK (4.12) – King (1991)				
Caginalp et al. (1998) – Asset rich & Dividend once	3.250	–	–	4;0;0
— Cash rich & Dividend once	0.000	–	–	3;0;0
Caginalp et al. (2001) – Asset rich & Call auction	10.600	–	–	5;0;0
— Asset rich & Call auction & Constant value	7.000	–	–	5;0;0
— Asset rich & Call auction & Dividend deferred	10.500	–	–	4;0;0
— Asset rich & Call auction & Dividend deferred & Open book	10.667	–	–	3;0;0
— Asset rich & Call auction & Open book	10.000	–	–	4;0;0
— Call auction	9.500	–	–	2;0;0
— Call auction & Cash rich	11.333	–	–	3;0;0
— Call auction & Cash rich & Constant value	6.167	–	–	6;0;0
— Call auction & Cash rich & Open book	11.250	–	–	4;0;0
— Call auction & Constant value	13.000	–	–	1;0;0
Davies (2006) – Baseline	6.667	–	–	3;0;0
— Increasing value	0.857	–	–	7;0;0
Hirota and Sunder (2007) – Dividend once & -certainty	5.000	–	–	1;0;0
— Dividend once & -certainty & -heterogeneity	5.000	–	–	4;0;0
— Dividend once & -certainty & -heterogeneity & short horizon	6.059	–	–	4;0;0
— Dividend once & short horizon	7.000	–	–	2;0;0
King (1991) – Information in each period	6.67	–	–	6;0;0
— Information in periods 1, 6, and 11	7.00	–	–	6;0;0
King et al. (1993) – Baseline	10.2	5.67	3.00	10;3;2
— Brokerage fees	10.0	6.0	–	2;3;0
— Equal endowments	10.0	–	–	4;0;0
— Informed insiders	13.0	10.0	–	1;1;0
— Informed insiders & short selling	13.0	4.00	–	1;2;0
— Limit price change rule	10.5	5.5	1.5	2;2;2
— Margin buying	8.00	2.00	–	1;1;0
— Short selling $\leq$ 2 units	9.50	5.80	3.67	4;5;3
— Short selling $\leq$ 2 units & margin buying	13.0	11.0	–	1;1;0
Noussair et al. (2001) – Constant value	8.000	–	–	8;0;0
This study - DO8	*8.000*	*5.250*	*3.000*	*4;4;1*
— DO5/10/15	*11.000*	*7.333*	–	*3;3;0*
DurationPS (4.13) – Porter and Smith (1995)				
Ackert and Church (2001) – Baseline	9.33	2.50	–	3;2;0
— Experienced business	8.00	–	–	2;0;0
— Experienced non-business	4.50	–	–	2;0;0
— Non-business	9.00	5.00	–	3;2;0
Caginalp et al. (1998) – Asset rich & Dividend once	2.500	–	–	4;0;0
— Cash rich & Dividend once	1.667	–	–	3;0;0
Caginalp et al. (2001) – Asset rich & Call auction	10.000	–	–	5;0;0
— Asset rich & Call auction & Constant value	6.000	–	–	5;0;0
— Asset rich & Call auction & Dividend deferred	10.500	–	–	4;0;0
— Asset rich & Call auction & Dividend deferred & Open book	9.333	–	–	3;0;0

(continued)

Table 9 (continued)

Measure Article / Treatment	Experience			
	None	Once	Twice	n
— Asset rich & Call auction & Open book	8.750	–	–	4;0;0
— Call auction	5.500	–	–	2;0;0
— Call auction & Cash rich	11.333	–	–	3;0;0
— Call auction & Cash rich & Constant value	4.333	–	–	6;0;0
— Call auction & Cash rich & Open book	4.000	–	–	4;0;0
— Call auction & Constant value	5.000	–	–	1;0;0
Corgnet et al. (2008) – Announcement high, preset	7.000	8.333	–	3;3;0
— Announcement low, preset	9.500	8.333	–	3;3;0
— Announcement high, random	11.67	–	–	3;3;0
— Announcement low, random	11.33	–	–	3;3;0
— Announcement true	8.33	–	–	3;3;0
— Baseline	9.167	8.000	–	3;3;0
Davies (2006) – Baseline	3.667	–	–	3;0;0
— Increasing value	1.143	–	–	7;0;0
Hirota and Sunder (2007) – Dividend once & -certainty	2.000	–	–	1;0;0
— Dividend once & -certainty & -heterogeneity	3.250	–	–	4;0;0
— Dividend once & -certainty & -heterogeneity & short horizon	3.195	–	–	4;0;0
— Dividend once & short horizon	4.000	–	–	2;0;0
Hussam et al. (2008) – Baseline	9.230[a]	6.479[a]	2.376[a]	34;13;8
— Cash rich & Dividend spread high	7.668[a]	5.001[a]	2.334[a]	3;3;3
— Rekindle	4.668[a]	–	–	3;0;0
Porter and Smith (1994) – Baseline	9.23	5.51	3.00	19;4;3
— Brokerage fees	10.00	4.00	–	2;3;0
— Dividend certainty	11.00	9.67	–	3;3;0
— Equal endowments	10.00	–	–	4;0;0
— Futures	10.00	5.50	–	3;2;0
— Informed insiders	13.00	6.00	–	2;3;0
— Limit price change rule	10.50	5.50	1.50	2;2;2
— Margin buying	8.00	2.00	–	2;1;0
— Short selling $\leq$ 2 units	9.50	5.80	3.67	4;5;3
Porter and Smith (1995) – Baseline	10.15	4.75	–	10;8;0
— Dividend certainty	11.00	9.7	–	3;3;0
— Futures & margin buying	10.00	5.5	–	3;2;0
— Margin buying	10.00	6.5	–	3;1;0
— Switch	–	–	4.5	0;0;2
Noussair et al. (2001) – Constant value	4.375	–	–	8;0;0
This study - DO8	*11.000*	*6.750*	*5.000*	*4;4;1*
— DO5/10/15	*11.000*	*7.000*	–	*3;3;0*
PositiveDurationACCD (4.14) – Ackert et al. (2006a)				
Ackert et al. (2006a) – Baseline & lottery asset	2.800	–	–	5;0;0
— Margin buying & lottery asset	4.500	–	–	4;0;0
— Short selling $\leq$ 5 units & lottery asset	1.500	–	–	4;0;0
Caginalp et al. (1998) – Asset rich & Dividend once	0.250	–	–	4;0;0
— Cash rich & Dividend once	0.667	–	–	3;0;0
Caginalp et al. (2001) – Asset rich & Call auction	4.400	–	–	5;0;0
— Asset rich & Call auction & Constant value	0.800	–	–	5;0;0
— Asset rich & Call auction & Dividend deferred	4.000	–	–	4;0;0

(continued)

Table 9 (continued)

Measure Article / Treatment	Experience None	 Once	 Twice	 n
— Asset rich & Call auction & Dividend deferred & Open book	2.000	–	–	3;0;0
— Asset rich & Call auction & Open book	3.750	–	–	4;0;0
— Call auction	2.000	–	–	2;0;0
— Call auction & Cash rich	8.667	–	–	3;0;0
— Call auction & Cash rich & Constant value	3.000	–	–	6;0;0
— Call auction & Cash rich & Open book	2.750	–	–	4;0;0
— Call auction & Constant value	4.000	–	–	1;0;0
Davies (2006) – Baseline	3.667	–	–	3;0;0
— Increasing value	0.143	–	–	7;0;0
Hirota and Sunder (2007) – Dividend once & -certainty	0.000	–	–	1;0;0
— Dividend once & -certainty & -heterogeneity	1.750	–	–	4;0;0
— Dividend once & -certainty & -heterogeneity & short horizon	2.445	–	–	4;0;0
— Dividend once & short horizon	3.000	–	–	2;0;0
This study – DO8	*10.750*	*6.750*	*5.000*	*4;4;1*
— DO5/10/15	*8.667*	*6.667*	–	*3;3;0*

[a]Results stem from a seemingly unrelated regression (SUR), not from averaging this measure directly over all applicable rounds

This table compares duration measure results reported in the literature for experimental asset markets. Results for the DO8 and DO5/10/15 treatments are printed in italics at the bottom of each block. The bubble measure names are followed by their formula number and the article they were first proposed in. All numbers reported in columns 2–4 are means over all rounds conducted with the stated treatment and level of experience. The last column lists the number of rounds the reported measures are the mean of, for each level of experience. When calculated specifically for this table, measures are listed with three digits of precision

$$\text{PositiveDurationACCD} = \max_t\left(m : \overline{P}_t - f_t < \overline{P}_{t+1} - f_{t+1} < \cdots < \overline{P}_{t+m} - f_{t+m}\right)$$
$$\text{s.t.} \quad \overline{P}_{t+1} > f_{t+1} \tag{4.14}$$

The difference between formulas (4.13) and (4.14) is that DurationPS would consider as part of a bubble periods which were followed by increases in the difference between price and fundamental value, even if they still constituted a negative bubble, in that the price remained below fundamental value. In the PositiveDurationACCD measure, bubbles consist only of periods where the mean transaction price actually was higher than the fundamental value.

Table 9 shows that the (temporal) length of the bubbles in the present experiment – both in terms of DurationK and DurationPS, and for inexperienced and experienced subjects – was relatively long, even though longer bubbles have been observed. The more precise measure of PositiveDurationACCD exacerbates the picture, since the present experiment displays the worst results of all studies reviewed.

4.1.2.4 Turnover Measure

In contrast to the previous categories containing several related measures, a single measure specification has unanimously been employed to describe the trading volume in Smith et al. (1988)-type experimental asset markets. It was first employed by King (1991), who measured the turnover by calculating a measure comprised of the total quantity of shares of the stock exchanged over the course of the experimental round, $\sum_t q_t$, divided by the number of shares outstanding:

$$\text{TurnoverK} = \frac{\sum_t q_t}{q} \tag{4.15}$$

The interpretation of the turnover in most of the markets in the literature is ambiguous. Due to the symmetric information structure, the no-trade theorem could be expected to apply for the case where all rational traders are risk-neutral. Even if subjects were heterogeneous with regard to their risk attitude, shares in this case should still only move from more to less risk-averse individuals (cp. the argument in Sect. 2.4.1.5). However, in most experiments, stocks tended to move back and forth between subjects, changing hands repeatedly over the course of the experiment – a fact that suggests that high turnover is indicative of an inefficient market. Smith et al. (2000) provided an alternative view, noting that if large numbers of trades occur around intrinsic value (as opposed to far away from intrinsic value, due to speculative activity), traders might infer that the market is highly competitive, which would inhibit price bubbles.[16]

Table 10 shows that the picture for the turnover is similar to that for the earlier bubble measures. The values for the present experiment are relatively high, but (except for the results for the single round with twice experienced subjects) do not constitute the peak observations, even excluding the results for Haruvy and Noussair (2006), which appear to be inflated.[17] It is particularly intriguing to note that increasing experience did not reduce the turnover in the present experiment as much as in other studies, even though the Active Participation Hypothesis would suggest that turnover in the spot market should be smaller in a setting where subjects have the additional option to divert themselves by trading in a derivative market. An explanation for this observation is not immediately apparent.

4.1.2.5 Dispersion Measures

For each treatment, King et al. (1993) calculated the mean variance of their prices (VarianceKSWV) over all sessions and trading periods. To justify the use of this

[16] Cp. Smith et al. (2000), p. 577.

[17] As noted at the bottom of Table 10, the Haruvy and Noussair (2006) turnover values according to their paper were calculated the same way as TurnoverK here (Cp. Haruvy and Noussair 2006, p. 1136).

Table 10 Turnover measure

Measure Article / Treatment	Experience None	 Once	 Twice	 n
TurnoverK (4.15) – King (1991)				
Ackert et al. (2002) – Baseline & lottery asset	2.75	–	–	5;0;0
— Margin buying & lottery asset	2.99	–	–	4;0;0
— Short selling $\leq$ 5 units & lottery asset	3.99	–	–	4;0;0
Ackert and Church (2001) – Baseline	2.45	1.05	–	3;2;0
— Experienced business	1.38	–	–	2;0;0
— Experienced non -business	0.80	–	–	2;0;0
— Non-business	2.02	0.85	–	3;2;0
Corgnet et al. (2008) – Announcement high, preset	4.14	2.65	–	3;3;0
— Announcement low, preset	4.31	2.53	–	3;3;0
— Announcement high, random	7.73[b]	6.29[b]	–	3;3;0
— Announcement low, random			–	3;3;0
— Announcement true	5.07	4.01	–	3;3;0
— Baseline	6.00	3.32	–	3;3;0
Davies (2006) – Baseline	4.998	–	–	3;0;0
— Increasing value	4.471	–	–	7;0;0
Haruvy et al. (2007) – Call auction	2.20	1.70	1.43	6;6;6
Haruvy and Noussair (2006) – Baseline	12.20	–	–	2;0;0
— Short selling $\leq$ 3 units	13.23	–	–	2;0;0
— Short selling $\leq$ 6 units	20.45[a]	5.34	–	3;3;0
— Short selling $\leq$ 6 units, cash times 10	22.86	–	–	2;0;0
— Short selling 100% cash reserve	21.09[a]	16.22	–	3;3;0
— Short selling 100% cash reserve, cash times 10	34.53	–	–	2;0;0
— Short selling 150% cash reserve	24.75	–	–	2;0;0
— Short selling flexible cash reserve	19.70	–	–	2;0;0
Hussam et al. (2008) – Baseline	3.074[c]	2.876[c]	1.217[c]	34;13;8
— Cash rich & Dividend spread high	2.645[c]	2.033[c]	1.511[c]	3;3;3
— Rekindle	2.095[c]	–	–	3;0;0
King (1991) – Information in each period	4.07	–	–	6;0;0
— Information in periods 1, 6, and 11	5.55	–	–	6;0;0
King et al. (1993) – Baseline	4.55	3.20	1.70	10;3;2
— Brokerage Fees	5.55	1.75	–	2;3;0
— Equal endowments	6.29	–	–	4;0;0
— Informed insiders	1.67	2.33	–	1;1;0
— Informed insiders & short selling	3.68	4.92	–	1;2;0
— Limit price change rule	4.84	2.22	1.89	2;2;2
— Margin buying	5.48	2.33	–	1;1;0
— Short selling $\leq$ 2 units	6.67	4.19	2.74	4;5;3
— Short selling $\leq$ 2 units & margin buying	3.60	6.89	–	1;1;0
Lei et al. (2001) – Baseline	5.640	–	–	4;0;0
— No speculation	0.847	–	–	3;0;0
— Two markets	0.782	–	–	6;0;0
— Two markets & No speculation	0.593	–	–	3;0;0
Noussair and Powell (2008) – Peak	7.800	3.468	2.456	5;5;5
— Valley	7.802	5.256	3.710	5;5;5
Noussair et al. (2001) – Constant value	4.19	–	–	8;0;0
Noussair and Tucker (2006) – Reverse futures	0.985	–	–	4;0;0
Porter and Smith (1994) – Baseline	5.79	3.00	1.60	19;4;3
— Brokerage fees	5.56	4.92	–	2;3;0
— Dividend certainty	8.84	2.71	–	3;3;0

(continued)

Table 10 (continued)

Measure Article / Treatment	Experience			
	None	Once	Twice	n
— Equal endowments	6.29	–	–	4;0;0
— Futures	6.85	2.63	–	3;2;0
— Informed insiders	2.68	4.05	–	2;3;0
— Limit price change rule	4.84	2.22	1.89	2;2;2
— Margin buying	5.48	2.33	–	2;1;0
— Short selling $\leq$ 2 units	6.67	4.19	1.74	4;5;3
Porter and Smith (1995) – Baseline	5.49	2.98	–	10;8
— Dividend certainty	8.85	2.71	–	3;3;0
— Futures & margin buying	6.85	2.63	–	3;2;0
— Margin buying	5.40	4.61	–	3;1;0
— Switch	–	–	2.59	0;0;2
Smith et al. (1988) – Baseline	5.634	3.501	1.519	7;6;3
— 1/3 experienced	n/a	2.111	–	n/a;1;0
— 2/3 experienced	n/a	3.222	–	n/a;1;0
— Dividend once	4.008	2.167	–	2;1;0
Smith et al. (2000) – Baseline	5.179	3.204	–	6;3;0
— Dividend mix	4.441	2.917	–	5;2;0
— Dividend once	5.459	3.223	–	8;2;0
Van Boening et al. (1993) – Call auction	3.075	1.680	2.055	2;2;2
— Symmetric dividend	7.030	5.285	3.965	2;2;2
This study - DO8	*6.313*	*4.219*	*4.042*	*4;4;1*
— DO5/10/15	*5.681*	*4.792*	–	*3;3;0*

[a]Calculated by combining Tables 2 and 3 of Haruvy and Noussair (2006)
[b]The turnover of Corgnet et al.'s (2008) two random message treatments was only reported in an aggregated form
[c]Results stem from a seemingly unrelated regression (SUR), not from averaging this measure directly over all applicable rounds
The Haruvy and Noussair (2006) turnover values appear inflated, but according to their paper were calculated the same way as TurnoverK here (Cp. Haruvy and Noussair 2006, p. 1136)
This table compares turnover measure results reported in the literature for experimental asset markets. Results for the DO8 and DO5/10/15 treatments are printed in italics at the bottom of each block. The bubble measure names are followed by their formula number and the article they were first proposed in. All numbers reported in columns 2–4 are means over all rounds conducted with the stated treatment and level of experience. The last column lists the number of rounds the reported measures are the mean of, for each level of experience. When calculated specifically for this table, measures are listed with three digits of precision

metric they wrote: "The volatility of prices in an experiment is measured by the variance of prices over the 15-period life of the asset. This is the most common measure of volatility for field data where fundamental value is not objectively defined." [18] For the results reported in Table 11, VarianceKSWV was calculated using the standard sample variance formula:

$$\text{VarianceKSWV} = \frac{1}{\sum_{t=1}^{T}(I_t)+1} \cdot \sum_{t=1}^{T} \sum_{i_t=1}^{I_t} (P_{i_t} - \overline{P}_t)^2 \qquad (4.16)$$

[18]King et al. (1993), p. 184.

Table 11 Dispersion measures

Measure Article / Treatment	Experience			
	None	Once	Twice	n
DispersionRatio (4.17) – This study				
This study - DO8	*0.308*	*0.252*	*0.245*	*4;4;1*
— DO5/10/15	*0.494*	*0.344*	–	*3;3;0*
VarianceKSWV – King et al. (1993)				
King et al. (1993) – Baseline	1.08	1.22	0.815	10;3;2
— Brokerage fees	0.526	0.615	–	2;3;0
— Equal endowments	2.22	–	–	4;0;0
— Informed insiders	0.57	1.72	–	1;1;0
— Informed insiders & short selling	0.192	1.23	–	1;2;0
— Limit price change rule	0.213	1.76	1.45	2;2;2
— Margin buying	7.96	6.15	–	1;1;0
— Short selling $\leq$ 2 units	2.20	1.79	1.46	4;5;3
— Short selling $\leq$ 2 units & margin buying	0.656	0.326	–	1;1;0
Smith et al. (2000) – Baseline	1.132	1.223	–	6;3;0
— Dividend mix	0.826	0.194	–	5;2;0
— Dividend once	0.701	0.351	–	8;2;0
This study - DO8	*1.338*	*4.693*	*1.860*	*4;4;1*
— DO5/10/15	*1.893*	*3.525*	–	*3;3;0*

This table compares dispersion measure results reported in the literature for experimental asset markets. Results for the DO8 and DO5/10/15 treatments are printed in italics at the bottom of each block. The bubble measure names are followed by their formula number and the article they were first proposed in. All numbers reported in columns 2–4 are means over all rounds conducted with the stated treatment and level of experience. The last column lists the number of rounds the reported measures are the mean of, for each level of experience. When calculated specifically for this table, measures are listed with three digits of precision

This measure has the advantage of being directly comparable to variance measures from empirical market studies. Nonetheless, it has a number of deficiencies when applied to the studies listed in this article, which are not being pointed out in the literature: One is that this dispersion measure increases with increasing nominal prices and price changes. Experiments using assets with lower fundamental values can be expected to yield lower absolute price changes. This results in a lower variance that is entirely due to the value range specified for the asset. Variance figures of this type cannot be compared to those of studies with assets of different fundamental values. The second major shortcoming of this measure is that it cannot be employed to compare studies like Smith et al. (1988), which employs an asset with decreasing fundamental value, with studies like Noussair et al. (2001), whose asset has constant fundamental value. Clearly, a comparison of the variances reported for studies using such heterogeneous fundamental value processes yields no meaningful results if it is conducted using the price variance as the medium of comparison.

The following new measure is related to the price variance, but redresses these shortcomings:

$$\text{DispersionRatio} = \frac{1}{T} \cdot \sum_{t=1}^{T} \frac{\hat{\sigma}_{P_{i_t}}}{\sigma_{f_t}}, \tag{4.17}$$

where $\hat{\sigma}_{P_{i_t}}$ is the sample standard deviation of transaction prices in period t, σ_{f_t} is the population standard deviation of the expectation of the ex post fundamental value of the asset in period t,[19] and the DispersionRatio thus measures the mean transaction price volatility relative to the volatility of the asset's fundamental value. A value of 1 signifies that the transaction prices in the experiment are on average exactly as volatile as the ex post fundamental value of the asset, with values smaller (larger) than unity signifying transaction prices that are less (more) volatile than this benchmark. This measure remedies both disadvantages of the variance measure discussed above – it is unaffected both by the absolute level of fundamental values, and by whether the fundamental value of the asset declines, remains constant, or increases over the course of the experiment.

Presenting a picture that is by now familiar, Table 11 shows that the sessions run in Graz and Klagenfurt exhibited high VarianceKSWV values, even though they remained below the outcomes for the *Margin buying* treatment of King et al. (1993). The DisperisonRatio is proposed here for the first time, and no article from the literature provided sufficient information to make it possible to calculate this measure for prior studies. For this reason, Table 11 lists results for this measure only for the present experiment, and does not allow for a comparison of findings across different studies and treatments (except for DO8 and DO5/10/15, which show the familiar pattern of higher efficiency in the former treatment).

4.1.3 *Subject Performance*

After the discussion of performance statistics at the transaction level in the previous section, the present section provides an analysis of the experiment conducted at the subject level. The total payout per subject in the DO8 (DO5/10/15) treatment ranged from € 0 to € 74.26 (€ 0 to € 121.48), with a mean of € 31.35 (€ 30.61).[20] Figure 8 shows a histogram of subjects' payouts in the two treatments. It also illustrates the positive skewness of the payoffs of 0.427 (1.895) in the DO8 (DO5/10/15) treatment.

[19] In experiments with 15-periods, possible (independent) dividend payments of 0, 8, 28, or 60 cents, and no terminal value, this is $\sqrt{(16-t) \cdot 0.25 \cdot \left[(0-24)^2 + (8-24)^2 + (28-24)^2 + (60-24)^2\right]}$ in period t.

[20] Of the 86 subjects, 5 would have attained a negative payout value (including the attendance fee) and received a payout of zero. Ironically, one subject would have received a payout of €-0.40 plus the €3 from the attendance fee, but seems to have forgotten about the latter, because she left after the experiment and did not step into the extra room to collect her payment of €2.60.

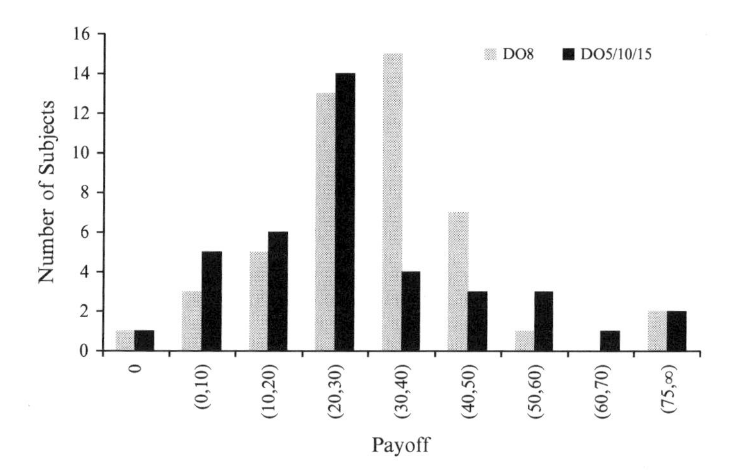

Fig. 8 Subject payoffs. Histogram of the total payouts in Euro to individual experimental subjects, including the attendance fee, by treatment

Because the subject population within a session is constant, observations from different periods within a session are not independent draws. Due to the different subject base, however, identical periods in different sessions can be considered independent draws. Using this statistical fact, empirical distributions of test statistics can be constructed.

The responses to the questionnaire completed by each subject after each experimental round were used to test for interrelation patterns between subject characteristics (mainly demographical items) and subject behavior and performance in the experiment. Table 12 contains both a translation of the questions and a summary of the answers. One test yielded the (non-)finding that there was no significant effect of subjects' education on their performance in the experiments (p-value of 0.545 in an OLS regression), a result that is in line with the literature overview of Camerer and Hogarth (1999), who similarly found little evidence that subjects with higher education performed better in experiments. The evidence also suggests that the subjects did not exhibit overconfidence when answering the questionnaire after the session. When asked to compare their performance to that of the average subject on a five-point scale (0 = much worse, 4 = much better), the median trader chose the value 2, while the mean player chose 1.875, a value that is significantly smaller than 2 at the 10%-level (p-value of 0.0871).

A more interesting result is the discovery that gender was significantly related to a number of subject characteristics, which are summarized in Table 13. On average, the male and female subjects were approximately of the same age (question 13 of Table 12) and had no significantly different levels of education (question 14). On the other hand, they differed with regard to their prior exposure to and knowledge about financial matters. Male subjects had more prior experience in financial topics, as evidenced by the higher percentage of male subjects who had previously traded stock (question 18) and options (question 19), and by males' higher mean answer on

Table 12 Translation of and answers to the questionnaire

Question	Scale	Mean				
		DO8	DO8x	DO8xx	DO5/10/15	DO5/10/15x
1. What was your strategy during the experiment?	Open/Text					
2. Did you change your strategy during the experiment and if yes, why?	Open/Text					
3. What do you believe was the strategy of the other subjects?	Open/Text					
4. Do you think that you acted rationally and that you were able to maximize your profit?	0–4	2.149	2.468	2.500	2.154	2.103
5. Did you at any time mistype something? If yes, please describe briefly when and how this mistake occurred!	Open/Text					
6. Did the option market help you gage how much you would be willing to pay for the stock?	0–4	1.362	1.404	1.083	1.513	1.333
7. Compared to the average player, how successful do you think you were?	0–4	1.787	2.021	2.250	1.872	1.692
8. What do you think is the reason for your having performed better/worse?	Open/Text					
9. How easy was it for you to understand the market mechanism and prepare a strategy?	0–4	2.596	2.702	2.750	2.564	2.795
10. How easy was it for you to understand the written and oral instructions?	0–4	3.170	3.404	3.333	3.282	3.462
11. How could the experiment be made easier to understand and/or what could be improved?	Open/Text					
12. Which studies have you completed or are currently enrolled in?	Open/Text					
13. Your age?	17–99	25.23	25.19	29.67	24.38	24.33
14. Your sex?	Female/Male	42.6[b]	42.6[b]	58.3[b]	28.2[b]	28.2[b]
15. Highest level of education completed?	[a]see below	2.160	2.160	2.160	1.359	1.359
16. Is this your first laboratory experiment?	Yes/No	66[c]	-	-	87[c]	-
17. How good would you say is your knowledge of finance and capital markets?	0–4	2.234	2.234	2.083	2.103	2.077
18. Have you traded stock in the past?	Yes/No	48.9[c]	48.9[c]	58.3[c]	41[c]	41.[c]
19. Have you traded options in the past?	Yes/No	12.8[c]	12.8[c]	16.7[c]	7.7[c]	7.7[c]
20. How often have you participated in a similar experiment involving trading of stock in a market?	Number	0.255	1.255	2.333	0.154	0.615
21. How often have you participated in a similar experiment involving trading of stock and options in a market?	Number	0.064	1.064	2.083	0.051	0.538

[a]Possible answers were: Elementary school (0), Apprenticeship (0), Secondary school (0), Grammar school (1), Bachelor (2), Master (3), Doctorate (4), Transformed into numbers as indicated by the values in brackets.

[b]Percentage of female subjects

[c]Percentage of "Yes" replies

The questions listed in this table were asked of all participants after each 15-period round. Questions scaled to the numbers 0–4 carried a caption that had one pole designated with a term similar to "Disagree completely" (corresponding to 0) and the other pole with a designation similar to "Agree completely" (corresponding to 4). Each "x" in the caption row corresponds to one session of prior experience for the subjects

Table 13 Characteristics of female and male subjects

Question	Scale	Mean		*p*-value
		Female (n=31)	Male (n=55)	
(4) Rational	0–4	2.174	2.287	0.126[c]
(6) Option market helped	0–4	1.419	1.291	0.643[c]
(7) Success	0–4	1.391	2.165	0.000[c]
(9) Understanding market	0–4	2.362	2.852	0.026[c]
(10) Understanding instructions	0–4	3.188	3.409	0.086[d]
(13) Age	17–99	26.0	24.6	1.000[c]
(15) Education	Nominal[a]	1.754	1.861	0.545[c]
(16) First experiment	Yes/No	58.1[b]	63.6[b]	0.649[c]
(17) Finance knowledge	0–4	1.78	2.39	0.000[c]
(18) Traded stock	Yes/No	29.0[b]	56.5[b]	0.000[c]
(19) Traded options	Yes/No	0.0[b]	17.4[b]	0.000[c]
(20) Previous stock experiment	Number	0.226	0.418	0.462[c]
(21) Previous option experiment	Number	0.129	0.236	0.394[c]
Relative result	Ratio	0.551	1.253	0.000[e]
Fraction of stock transactions	Percentage	7.511	8.537	0.181[e]
Fraction of option transactions	Percentage	6.210	9.317	0.002[e]

[a]Possible answers were: Elementary school (0), Apprenticeship (0), Secondary school (0), Grammar school (1), Bachelor (2), Master (3), Doctorate (4), transformed into numbers as indicated by the values in brackets
[b]Percentage of "Yes" replies
[c]The *p*-value stems from Fisher's exact test
[d]Since the median answer to this question was 4 for both sexes, no Fisher exact test *p*-value could be calculated. The *p*-value therefore stems from a two-sample Mann-Whitney test
[e]The *p*-value stems from an OLS regression with robust standard errors
Comparison of mean questionnaire answers and performance in the experiment, conditional on subject sex. The question numbers in column 1 correspond to the numbers in Table 12. The last column lists the two-sided *p*-values for a hypothesis of equal means. Relative result is a subject's monetary result net of the attendance fee, divided by the mean result of all subjects in the same round. The fraction of stock (option) transactions measure the fraction of all transactions within a round conducted by an average subject of a given sex

question 17 of Table 12, inquiring after their financial knowledge. Male subjects (maybe because of their better finance knowledge) also seemed to have an easier time understanding the instructions (question 10), and understanding the market mechanism and forming a strategy (question 9), which made them more confident with regard to their success in the market (question 7). These indications that the male subjects in this experiment seemed to be more confident with regard to their knowledge about financial affairs were supported by the outcomes. Male subjects traded significantly more frequently in the option market and seemed to also trade slightly more frequently in the stock market (as measured by the fraction of all transactions within a round in the respective markets conducted by the average subject of each sex). Finally, normalizing each subject's relative payoff (excluding the attendance fee) by the mean payoffs of the experimental session they participated in (the mean cohort payoff) yields a measure of relative performance (designated

"Relative result" in Table 13).[21] A linear regression of this performance measure on subject gender found higher values for male subjects (factor loading: 0.496). Even controlling for factors like the previous trading experience, number of transactions in stock and option markets, etc., this result was highly significant (p-value << 0.01).

The reason for this marked difference between subjects of the two sexes remains unclear. While some sources report that women are less overconfident and trade (i.e., churn) less than men, the author is aware of no studies finding that women are on average less well-informed about financial matters.[22] Interestingly, women in the experiments conducted for this book were more confident with regard to their rationality and the extent of their profit-maximizing behavior in the experiment (question 4). Unfortunately, this issue of the difference between the two sexes cannot be probed in more detail, since such differences were not initially at the heart of this research effort. Therefore, no analyzes specific to this research question were planned beforehand, which precluded a detailed ex post analysis.

4.2 Interpretation of Behavioral Regularities

4.2.1 The Hypothesis of Observational Belief-Adaptation

Literature is strewn with the wreckage of men who have minded beyond reason the opinion of others.

Virginia Woolf, A Room of One's Own, 1929

The experiment presented in this book was modeled after the seminal paper of Smith et al. (1988), who found that although the possible dividend draws were common (symmetric) information and every subject had all the information required to derive the fundamental value of the stock in every period, there was a persistent pattern across their inexperienced subjects: The stock price started out below its fundamental value in period 1. Over the course of the experiment, the stock price then rose above its fundamental value, creating a bubble. During the final periods, the price eventually crashed down to levels close to its fundamental (intrinsic dividend) value. This pattern can also be observed in the results of the current experiment.

As evidenced by the bubble measures in Sect. 4.1.2, price processes in the experiment follow paths similar to those reported for the baseline series in the literature. Over the course of the experiment, prices rose from close to the dividend holding value to levels significantly exceeding this fundamental benchmark, and

[21] The absolute performance of subjects does not only depend on their actions, but also on the random dividend draws in their experimental session. This effect is filtered out by the normalization described above.

[22] Malkiel even recommends the solicitation of female advice when making investment decisions. However, he does not comment on how well-informed women are on average (Cp. Malkiel et al. (2005), p. 130).

then crashed back to levels close to the stock's intrinsic value during the later periods.[23] The market's behavior in the phase of increasing prices can be described with the words "irrational exuberance," a phrase that was made famous by the former chairman of the board of the U.S. Federal Reserve, Alan Greenspan.[24] It describes a situation where the market participants seem to believe that prices will continue to rise indefinitely, or at least until they find another trader to sell their asset holdings to.[25]

Many scholars conjectured that this pattern may have been caused by the following mechanism:[26] In the early periods, subjects – being yet inexperienced in this market – trade the stock at a discount from its expected value in an expression of risk aversion.[27] Gaining experience and confidence over time, prices increase (both in absolute terms and relative to fundamental value), creating an upward trend. Observing this tendency of positive price changes, and noting that the direction of price changes is independent of the direction of changes in fundamental value, subjects then extrapolate and prices increase to levels exceeding the fundamental dividend value. During this period, buyers are conjectured to hope to gain from either selling the stock for an even higher price later on, or from selling it for about the same price (or even a smaller price) while pocketing the intervening dividend payments. Finally, as the last period approaches, traders' subjectively perceived probability of being able to sell their stock at the current price level decreases enough to induce large-scale attempts to sell stock, which naturally – since most traders followed a similar strategy – no longer finds buyers at the inflated price. This finally causes the last piece of the observed price pattern to fall into place: Stock prices drop precipitously, at very low volumes of trade. Porter and Smith (1995) gave the following vivid account of this assumed process:[28]

> "Think of a story such as the following: given their disparate initial portfolios and attitudes toward risk, those most eager to balance their portfolios in line with their risk attitude trade at discount prices that provide a premium to the more risk-averse buyers. At these low initial prices, other subjects start to execute arbitrage purchases. The resulting price increase sets up expectations of capital gains from a further rise in prices. Self-fulfilling capital gains expectations then drive the bubble to ever higher prices until near the end when it becomes transparent that a correction is in order."

---

[23] The price deviations were in fact so strong that transaction prices were found to be significantly larger (p-value from an unparametric Wilcoxon matched-pairs signed-ranks test: 0.0000) than the fundamental value over the entire experiment (as well as by round), not only over the bubble phase within the rounds.

[24] Cp. Greenspan (1996).

[25] Cp. Miller (2002), p. 1.

[26] Cp. e.g., Smith et al. (1988), p. 1149.

[27] Caginalp et al. (2000b) noted that it is a common characteristic of first-period trading that buyers tend to have low share endowments, while sellers are more likely to have high share endowments. This would suggest that the reason for early trades is a wish to rebalance skewed cash and asset endowments by risk-averse subjects.

[28] Porter and Smith (1995), p. 514–515.

Note that in the above explanation, trading at prices above fundamental value is not necessarily a sign of irrationality, but requires only that rationality of all traders not be common knowledge (see Sect. 2.4.1.5 on no-trade equilibria). This point is of considerable importance, as the literature up until the beginning of the current century attempted to explain the bubble-and-crash pattern using a framework of rational agents lacking common knowledge of rationality. Smith et al. (1988) for example suggested that bubbles can be caused by rational traders doubting the rationality of other traders, which leads them to speculate in the belief that the future offers opportunities for capital gains.[29] In other words, bubbles are possible in a world where all traders are rational, but where they lack common knowledge of this fact. Such an argument was also advanced by Plott (1991), who unambiguously stated his belief that agents are rational and learn about others' rationality over time, explaining the observation of disappearing irrationality in agents' behavior.[30] Lei et al. (2001) refer to this hypothesis as "the speculative hypothesis." They then write:[31]

> "To see how a bubble and crash might come about if it is not common knowledge that traders are rational, consider a rational trader who believes that there may be "irrational" traders in the market, who are willing to make purchases at very high prices. The rational trader might make a purchase at a price greater than the fundamental value, believing that he will be able to realize a capital gain by reselling at an even higher price, either to an irrational trader or to a trader who also plans on reselling. Thus trading prices may be much higher than the fundamental value when the end of the time horizon is sufficiently far in the future, even when all agents are rational. However, as the end of the time horizon approaches, the probability of realizing a capital gain on a purchase declines, the incentive to speculate is reduced, and the price falls (crashes) to the fundamental value. It need not be the case that irrational traders actually exist, but only that their existence be believed to be possible."

However, Lei et al. (2001) then went on to refute this explanation by showing that the rationality assumption was violated and that subjects' actions in this type of market are at least partly founded in irrationality and myopia (see Sect. 2.4.4.3 for a more detailed account of their experiment and the findings thereof).

Moving on from this verdict on the causes underlying the price pattern with inexperienced subjects, the question arises what causes can be identified for the decline of the observed bubbles in rounds with experienced subjects. In the past 20 years, the bubble-and-crash observation has been replicated numerous times, with a wide array of variations in market mechanics, subject pool, dividend structure, etc., and has been found to be remarkably robust. As mentioned before, the only variable that has been known to reliably lead to a disappearance of the observed bubble was experience.[32] Subjects who have played the same

---

[29] Cp. Smith et al. (1988), p. 1148.

[30] Cp. Plott (1991), p. 916–917.

[31] Lei et al. (2001), p. 832.

[32] See footnote 55 on p. 40 for a qualification of this statement.

experiment once or twice before usually produce a price series that follows the fundamental price series significantly more closely, with bubbles frequently vanishing entirely by the second repetition. This is also true for the experiment conducted for this book, since experience has been shown to have improved (i.e., lowered) all measures of the extent of the price bubble (see Tables 7 through 11 in Sect. 4.1.1 above).

It is at this point – in the description of the learning process in repeat rounds – that the literature departs from the observations of this study (as well as from those of Haruvy et al. (2007), which will be discussed in the next section). The general sentiment in the literature appears to be that individuals become more rational in the context of this type of experimental market, both over the periods within an experimental round and over repetitions of entire rounds. Lei, Noussair and Plott for example wrote:[33]

> "Over the course of the experiment, some traders come to realize that there is the possibility of irrational behavior on the part of other traders. This realization promotes speculation. Later, experience and practice reduce subject confusion and remove the irrationality of market participants. Once the irrationality has been removed, the new information about the change in the environment must be transmitted to the market. If our view is correct, that transmission takes the form of a crash. That is, the market crash is the vehicle whereby the newly established rationality of market participants becomes common knowledge.
>
> The duration of a bubble in [our treatment that precludes speculation] measures the length of time that irrationality is present among market participants. This is because bubbles in [this treatment] must indicate actual irrationality, not the lack of common knowledge of rationality. Because there is no evidence that the length of time the bubbles last is any shorter in [the treatment without the possibility of speculation] than in the other treatments in which speculation is possible, the period of time in which rationality is present but is not common knowledge is likely to be at most very short. Therefore, price crashes in markets with resale appear also to correspond to the beginning of the existence of rationality itself among all active market participants, rather than merely the beginning of common knowledge of rationality already present."

Such a model of agents' learning implies a convergence to rational expectations as defined by Muth (1961), in that agents learn to act in a way such that outcomes reinforce their decisions, where the outcomes can be derived from a rational model.[34] In other words, they step-by-step "discover" the rational fundamental market price.

This conflicts with the observations made over the course of this research project. Findings from the DO8 and DO5/10/15 experiments suggest that experience and practice reduce irrational actions by market participants in this particular type of market, but provide no evidence of a reduction of subject confusion and of the irrationality of market participants. In other words, there is no evidence that

[33] Lei et al. (2001), pp. 858–859.

[34] See Sect. 2.4.1 for a discussion of different models of expectation formation.

subjects who at the beginning of the experiment had not grasped the market logic of the asset market and the stock's fundamental value process had done so by its end. Naïve traders[35] over time learnt to trade at prices that coincided with the stock's fundamental value, but their answers in the questionnaire and statements in conversations with the experimenter after the sessions did not indicate that they were aware of this congruence of prices with fundamental value. In short, rather than lending themselves to explanation in terms of Muth rationality, their actions were more compatible with Nash rationality – their outcomes reinforced their actions, but there was no understanding or moving toward the underlying theory involved.

In a statement that is consistent with these conjectures, Porter and Smith (1995) concluded that common information about the fundamental value of the stock is not a sufficient condition for common expectations.[36] Due to behavioral uncertainty and irrationality, common expectations and convergence to fundamental value only come about through experience, not through logic applied to common information. This conjecture shall be referred to as the *Hypothesis of Observational Belief-Adaptation*:

> *Common expectations and convergence to fundamental value in Smith et al. (1988)-type experiments are due to learning from observation, not due to logic applied to common information.*

Note that this hypothesis is consistent with the general findings of Lei et al. (2001) that bubbles are not (solely) due to subjects speculating on being able to sell a stock that they buy at a high price (relative to fundamental value) today for an even higher price in the future, but are (at least in part) caused by outright irrationality.

Figure 9 below plots the deviation of the price paths of mean transaction prices from the fundamental prices induced by the expected subsequent dividend payments, for each treatment and period.[37] The following observations show that the empirical findings of this book support a mechanism of learning-by-observation as described above: Subjects indicated in both their answers to the post-period questionnaire and in conversations with the experimenter during and after the

---

[35] See Sect. 4.2.3 later in the text for a characterization of a "naïve trader". For the understanding of the present paragraph, the exact meaning of the term as used in this book is unnecessary.

[36] Cp. Porter and Smith (1995), p. 512. However, in the next sentence they went on to explain the bubble phenomenon by the familiar explanation of it being caused by behavioral uncertainty and the lack of common knowledge of rationality.

[37] Using feedback from subjects during the experiment and from the questionnaires, some outliers that could be clearly identified to have been caused by errors in data entry were identified and removed. Specifically, one outlier was removed from the stock price series of period 6 in the third session of experiment 1, one from period 8 in the first session of experiment 4, one from the first (period 9) and two from the second session (both period 15) of the fifth experiment and one each from the two sessions (period 7 and period 1, respectively) of the sixth experiment. This corresponds to 0.467 such outliers per session, or 0.031 per period.

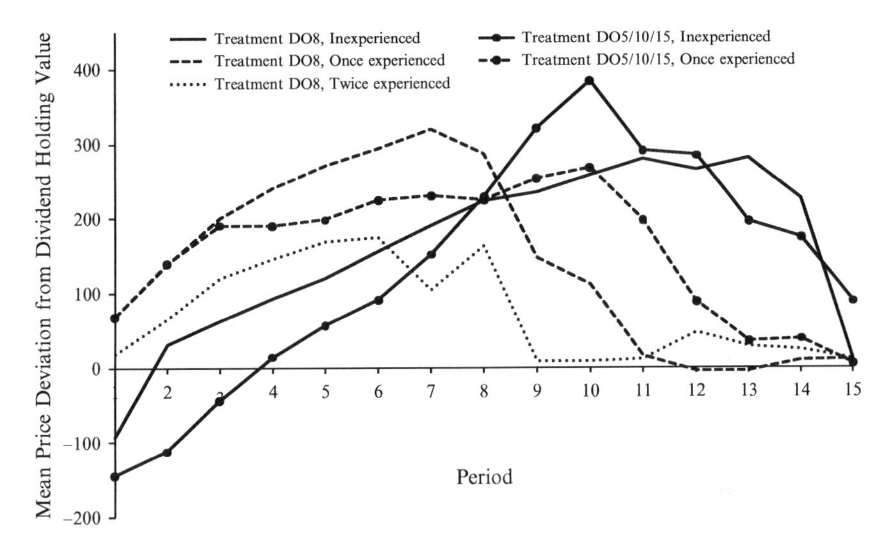

Fig. 9 Mean stock price deviation from dividend holding value. The five plots show, for each treatment and three levels of experience, the mean over all sessions of the mean stock price deviation (in cents) from fundamental value per period. Inexperienced subjects have never participated in an experiment of this type, once experienced subjects are the same individuals in the second round, and twice experienced subjects are the same individuals in the third round. There were no periods without stock transactions

experimental sessions that they were surprised by the paths prices take during the first round (the two solid lines in Fig. 9), which in all sessions conformed closely to the typical bubble-and-crash pattern documented in most experiments since Smith et al. (1988). Consistent with the Hypothesis of Observational Belief-Adaptation, subjects then seemed to "learn to bubble and crash" from this first round. More specifically, subjects learnt that the round started with strongly positive price developments, followed by a price downturn toward the end. By anticipating this (by then familiar) pattern, they strengthened the pattern of large positive price changes in the early periods and hastened the subsequent return of prices to levels near fundamental value.[38] As Fig. 9 shows, the mean price paths of the once (twice) experienced treatments up to period 7 (6) were strictly above those of the inexperienced treatments. Also, in the experienced treatments, prices were

[38] Some subjects noted in the questionnaires and in conversations with the experimenter after a round that they attempted to engage in a scheme of insuring their portfolios using synthetic puts, i.e., selling stock upon price declines. As Grossman (1988) argues, such a scheme can only work if the stock price is independent of the amount of money protected by such synthetic puts (p. 277). Naturally, this condition is not fulfilled in the small experimental market employed here. These strategies are therefore bound to fail, since the sales attempts induced by price declines tended to lead to further price declines without or with very little trade.

closer to dividend value than in the inexperienced treatments from period 11 onwards.[39]

The same pattern can be observed in Fig. 1 of Porter and Smith (1995), and even Smith et al. (1988) noted that their sessions 26 and 28x (both using the same subjects) conformed to this pattern[40] – evidence consistent with the Hypothesis of Observational Belief-Adaptation. If, by playing the first round, subjects had improved their understanding of the market and recognized that prices at the end of the session tended toward fundamental value (as conjectured in the literature), one would assume that the crash should occur earlier – as it did – and that the run-up to the bubble should have been dampened – which it was not. Finally, the Hypothesis of Observational Belief-Adaptation also predicts why increasing experience led to more efficient prices. Subjects who observe a crash in the first round usually anticipate this pattern, which leads to an earlier crash in the second round. By the third round, subjects have learnt that the moment at which prices start to return to fundamental values moves forward through time, and in many experiments they no longer even produce a bubble that could then lead to a crash.

Note that the Hypothesis of Observational Belief-Adaptation makes no predictions about the first round inexperienced subjects participate in, since it relies on observations in *earlier* rounds to predict behavior in later rounds. It thus offers no new evidence regarding the question of why bubbles form in the first place. However, it can be combined with Porter and Smith's (1995) explanation of low initial prices due to risk aversion, increasing prices due to a movement toward fundamental value, continuing price increases due to trend extrapolation, and a crash due to the decreasing probability of possible resale at inflated prices. Still, this

---

[39]Smith et al. (1988) ran experimental markets (30xsf and 39xsf) to show that the bubble-and-crash pattern did not originate (only) in subjects' inexperience. They first let subjects gain experience in an experiment where capital gains or losses were impossible across trading periods, because endowments were re-initialized at the beginning of each period. They then found that these subjects still produced the familiar bubble-and-crash patterns in subsequent experiments in *Baseline* markets, concluding that living through a bubble-and-crash pattern while inexperienced was not a precondition for this pattern to reappear in later periods. (Note that this test does not touch on the observation made here that – conditional on *having* observed a bubble-and-crash pattern in their first experimental round – subjects produce earlier and stronger bubbles as well as faster crashes in subsequent rounds.) Together with the findings of Lei et al. (2001) on the occurrence of bubbles in settings precluding speculation, it can be concluded that a horizon of more than one period (Smith et al. (1988), p. 1133, report observations of bubbles in rounds as short as three periods) is a necessary condition for bubbles and crashes to arise, while the possibility of speculation is not.
For completeness, it should also be mentioned that Smith et al. (1988) conducted an experiment with a horizon of 30 periods and found that this treatment did not produce an observable bubble-and-crash pattern, even if analytical measures detected a number of small anomalous price movements.

[40]The author only became aware of this observation in Smith et al. (1988) after he had identified this effect in the data of his own experiments – both because of observations during the experiments, but especially because of replies to the questionnaires.

explanation harbors an inconsistency in that subjects are assumed to be rational, yet extrapolate the increasing prices beyond the level of fundamental value. This inconsistency can be resolved by recognizing the findings of Lei et al. (2001) and abandoning the assumption of subject rationality.

In fact, the market structure itself may be responsible for the movement toward choices mimicking those of rational agents. Gode and Sunder (1993), in their zero-intelligence trader simulation, concluded with regard to the double auction market institution:[41]

> "The primary cause of the high allocative efficiency of double auctions is the market discipline imposed on traders; learning, intelligence, or profit motivation is not necessary.
>
> [...] Adam Smith's invisible hand may be more powerful than some may have thought: when embodied in market mechanisms such as a double auction, it may generate aggregate rationality not only from individual rationality but also from individual irrationality."

While the Smith et al. (1988) and subsequent experiments are not consistent with high market efficiency without learning, once subjects have had the chance to learn about the market, Gode and Sunder's second statement is a good description of the observations. In a similar vein, Stanley (1997) observed that even in irrational markets, some backward induction is likely to exist, which serves to anchor bubbles to the known future values. He wrote, that "Even when investors hold irrational expectations, experience is likely to make them painfully aware of the certain terminal value,"[42] and then noted that such anchoring of expectations could explain the findings of Smith et al. (1988) and Van Boening et al. (1993) that experience reduces and finally eliminates bubbles.

4.2.2 The Haruvy, Lahav and Noussair (2007) Study

Haruvy et al. (2007) identified evidence supporting a similar expectation formation process as that postulated by the Hypothesis of Observational Belief-Adaptation. More specifically, subjects in their experiment failed to predict a crash in round one (they initially predicted constant transaction prices over time) and overestimated the time remaining before the price peaked in rounds two to four.[43] In the latter three rounds, they also consistently overestimated the magnitude of the bubble in future periods of the current round. By soliciting forecasts for the mean period price in all future periods from their subjects, Haruvy et al. also showed that a simple

---

[41] Gode and Sunder (1993), pp. 134 and 136.

[42] Stanley (1997), p. 615.

[43] The peak price period in their study was defined as the period in which the highest (absolute) price occurred, with the last such period selected in the case of ties. Cp. Haruvy et al. (2007), p. 1906.

model of adaptive expectations was able to outpredict an expectation formation model assuming rational expectations. Summarizing these findings, they succinctly noted:[44]

> "This [...], coupled with the fact that bubbles decline in magnitude as the market is repeated, suggests that prices converge toward fundamentals ahead of beliefs. [...] we [...] analyze the determinants of expectations of price patterns in the market in detail, and find that expectations are primarily adaptive. They reflect anticipation of a continuation of previous trends from one period to the next, as well as from one market to the next [...]. Traders employ profitable strategies given their adaptive expectations, increasing net market demand when they expect prices to rise, while increasing net supply when they believe that a market peak and downturn is approaching. [...] The trading behavior just described reduces the size of bubbles and induces earlier price peaks with repetition of the market, moving the time series of transaction prices closer to fundamentals. After prices and expectations have converged to fundamentals, [...] expectations have become accurate predictors of future prices."

They illustrated the forward movement through time of the period containing the peak price in their experimental rounds in their Fig. 2. Figure 10 plots the same data for the experiment reported in this text. Not plotted is the peak price period for the third round of the present experiment, as only one such round was played. In this third round, prices peaked in the third period.

The mean peak price periods in the Haruvy et al. (2007) experiment (the experiment reported in this book) were 12.2 (8.4) in the first round, 6.3 (5.4) in the second, 3.5 (3) in round three, and 1.8 in the fourth round.[45]

Haruvy et al. (2007) then went on to estimate a simple linear regression model of prices, using transformations of past prices and subjects' predictions of future prices as the regressors.[46] They found that both a variable measuring the same-period price change in previous rounds and one measuring the trend in price changes within the round that was subject of the prediction were highly significant and together yielded R^2-values above 0.5 for all and above 0.7 for rounds two to four. A benchmark model based on fundamental values yielded R^2-measures of no more than 0.5 for all and less than 0.4 for rounds one to three. Using a similar regression, they then showed that in rounds two to four, subjects systematically predicted a longer time until the peak price period than was ex post actually the case. This is consistent with the Hypothesis of Observational Belief-Adaptation, where subjects expect a crash in a future period and – wanting to profit from it – wish to anticipate it. However, their sales offers in anticipation of the crash then

[44] Cp. Haruvy et al. (2007), p. 1905.

[45] These numbers were calculated from Haruvy et al. (2007), Fig. 2, p. 1907. On the same page, the authors reported different numbers, which are in conflict with the information in their figure. Upon inquiry, Ernan Haruvy confirmed to the author that the reported numbers in their article were in error and that the numbers listed above are correct. The difference is caused by differing treatments of ties in peak prices.

[46] Such a regression can unfortunately not be performed for this study's data, since no future price predictions were solicited from the experimental subjects.

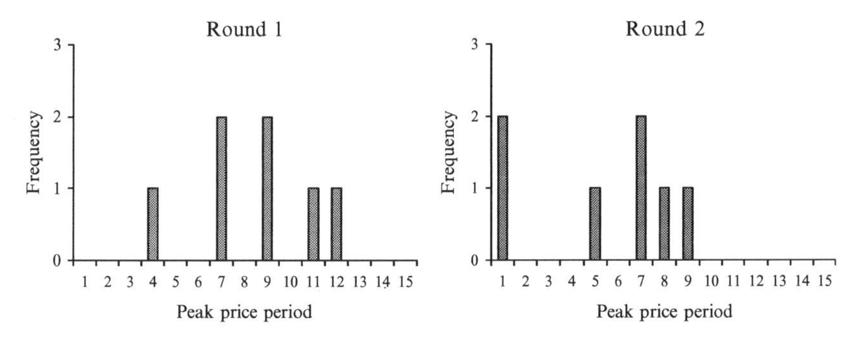

Fig. 10 Actual peak price period in each round. The figure shows the peak price period in all sessions of rounds 1 and 2, defined as the period in which the highest (absolute) price was observed in the stock market, with the last such period selected in the case of ties

trigger the crash early, thereby robbing them of capital gains and moving the peak price period forward through time.

Haruvy et al. summarized:[47]

> "Prediction biases during a bubble appear to occur because, while individuals base their predictions on history, they also optimize their trading behavior accordingly. Individuals attempt to reduce purchases and to increase sales when they anticipate that a price peak is imminent. The effect of this behavior is to cause deviations of prices from traders' predictions, to attenuate bubbles, and to make market peaks occur earlier than they did in markets the same individuals participated in previously. Because expectations are adaptive, the ever-smaller bubbles and earlier peak price periods influence, in turn, the predictions in the next market. The final result of this process is that bubble magnitudes converge toward zero and the peak price period converges toward period 1, in accordance with fundamental value pricing. By the fourth market in which a group of traders participates, prices track fundamental values closely. Convergence of asset markets to fundamental values in our markets thus appears to occur because traders use trading strategies that are profitable given their expectations, which are in turn based on history. That is, adaptive expectations, coupled with profit maximization, characterize a dynamic process of convergence toward fundamental pricing."

Note that, while the adaptive expectations literature is a well-established part of behavioral economics, its application to Smith et al. (1988)-type experimental markets *without* any assumption of rationality is a novel contribution pioneered by Haruvy et al. (2007) and the present text.[48]

---

[47] Haruvy et al. (2007), p. 1918.

[48] After having conducted the last search for new articles relevant for this book in November 2007, the author was made aware of this article by his discussant at the Spring Meeting of Young Economists 2008 in Lille, France, in February 2008. The results of this text were then updated to give due credit to the findings of Haruvy et al. (2007).

4.2.3 Bounded Rationality and Irrationality

> *People who trade on noise are willing to trade even though from an objective point of view they would be better off not trading. Perhaps they think the noise they are trading on is information. Or perhaps they just like to trade.*
>
> Fisher Black (1986)

The hypothesis explored in this study was that the opportunity of trading in an option market (and of observing the prices of trades in such a market) would lead to more efficient price vectors in the underlying spot market. To gauge why the hypothesis was found to be rejected (as it clearly was, considering the bubble measure results reported in Sect. 4.1.1), it might be instructive to investigate the structure of trading that took place in the two markets.

The mean trading volume of the stock market, per treatment and period, is plotted in Fig. 11 below. If turnover is regarded as a bubble measure, experience clearly led to a more efficient market. In 25 out of 30 cases, the mean turnover decreased from the inexperienced treatment to the once experienced treatment.[49] Assuming a binomial distribution with a probability of success of 0.5, a random draw would yield this result with a probability of less than 10^{-4}. Interestingly, all four cases where mean turnover increased from the inexperienced to the experienced treatment occurred in the DO5/10/15 treatment.[50] Contrary to the conjecture that more option maturity times would increase the efficiency of the market, this measure indicates that the opposite was the case.

Taken together with the more general observation that the option market did not seem to have increased the efficiency of experimental asset market prices, this effect requires an inquiry into its causes. The beginning of such an explanation may be found in the answers to the post-round questionnaire, which are summarized in Table 12 in Sect. 4.1.3. The sixth question asked whether subjects felt that the option market had helped them in determining how much they would be willing to pay for the stock. The mean answer on a scale from zero to four, where zero corresponded to "Not at all" and four to "Very much," was 1.362 (1.404) for the inexperienced (once experienced) subjects in the DO8 treatment and 1.513 (1.333) for the DO5/10/15 treatment. These relatively low numbers are an indication that subjects did not find the option market to be very useful in helping them form expectations about future prices.[51]

---

[49] The case for the twice experienced session is not as clear-cut, a result that is not unlikely to be spurious in light of the sample size of one.

[50] The one missing case stems from period 9 of the DO5/10/15 treatment, where turnover remained constant from the inexperienced to the experienced rounds.

[51] The results from the twice experienced subjects are not mentioned here since they stem from only one experiment, which – drawing from a pool of faculty members of the department of social and economic sciences of the University of Graz – had an atypically large number of subjects with sophisticated financial know-how.

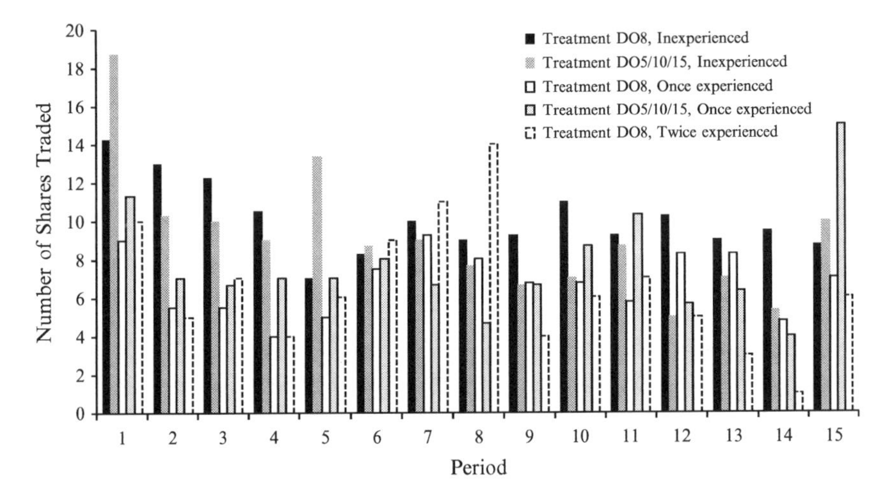

Fig. 11 Mean trading volume. The bars show, for each treatment and three levels of experience, the mean trading volume over all sessions. Inexperienced subjects have never before participated in an experiment of this type, once experienced subjects are the same individuals in the second trading round (i.e., they have by then participated once in such an experiment), and twice experienced subjects are the same individuals in the third trading round

The first intuition might be that the instructions regarding the option market were unclear. This suspicion is refuted by the relatively high mean answer values on question nine, inquiring how easy it had been for subjects to understand the market mechanism and to form a strategy, which were 2.596 (2.702) for inexperienced (once experienced) subjects in DO8, and 2.564 (2.795) for DO5/10/15, and the mean answers on question ten, asking how easy it had been for subjects to understand the written and oral instructions, which were 3.170 (3.404) for DO8 and 3.282 (3.462) for DO5/10/15.

A second strategy in trying to explain why the option market did not aid subjects in their expectation formation is to try to identify groups of subjects differing from each other with regard to their trading strategies and levels of understanding of the market. To this end, a detailed review of the questionnaires and the solicitation of personal feedback in conversation with subjects after the experiments were conducted. It was found that traders can be tentatively assigned to two broad groups, which will be designated *naïve* and *rational*. Naïve traders formed price expectations based on the current round's market prices and market prices in earlier rounds, in line with the Hypothesis of Observational Belief-Adaptation. They did not condition their expectation formation on fundamental information about the dividend value of the stock, but learnt from their observations of market prices. Rational traders, on the other hand, had fully understood the market mechanism and fundamental value process of the stock and initially traded based on this prior information. These findings can be condensed to the following conjectures:

Traders can be assigned to the two groups of (1) *naïve traders and* (2) *rational traders. They differ with regard to their grasp of the market mechanics, of the stock's fundamental value process, and with respect to their expectations formation.*

Naïve traders follow the Hypothesis of Observational Belief-Adaptation. They trade based on prices in the current and in past rounds, but do not condition on the fundamental value.

Rational traders understand the market mechanics and the fundamental value process. In the presence of naïve traders, they initially trade based on rational expectations, but soon speculate in the expectation of being able to invert their transactions in the future, reaping capital gains or dividend income in the intervening time.

These two types of traders usually also displayed two characteristics Shiller (2003) had described for his groups of "smart money" and "irrational traders": First, investors tended to stay in their group throughout the experiment, meaning that naïve (rational) traders tended to stay naïve (rational), and did not develop into rational (naïve) traders as the experiment progressed, which is an important part of the Hypothesis of Observational Belief-Adaptation and was already alluded to in Sect. 4.2.1 above. Second, as prices rose above fundamental values, rational traders could often be observed selling their stock, a behavioral pattern that is often identified in the literature as causing prices to remain at or return to fundamental values. However, both in the experiments conducted for this book and in the argument of Shiller (2003), all that happened was that rational traders ran out of stock and naïve traders kept transacting at prices far exceeding fundamental values. Compounding this effect was that, in later periods and sessions, rational traders sometimes learnt from the observed market prices and at times speculated on the actions of their naïve counterparts, by purchasing shares above the fundamental value in the expectation that they would be able to earn capital gains and dividends when selling them later at similar or higher prices.[52]

This observation of seemingly naïve trading by rational subjects resembles closely the phenomenon of destabilizing rational speculation from the De Long et al. (1990) model, which was briefly summarized in Sect. 2.1.3. Furthermore, Smith (1985) identified similar groups of traders in real-world financial markets:

"I find it necessary, if not entirely satisfactory in terms of seeking a universal theory, to accept the idea that some people just simply like to gamble (ancient hunter cultures did it) and that it has commodity value, or perhaps that some people have 'pathological' expectations, whether it is roulette, grain futures or stock investment."[53]

"If in all markets with uncertainty there is a subclass of participants with these 'irrational' characteristics, this lowers the insurance cost of hedging and lowers the cost of capital

[52] Since these types of traders are not the input for a theoretical model but rather a deduction from empirical observations, subjects do not perfectly conform to these stylized characteristics. Nonetheless, these cognitive constructs may constitute a conceptual simplification of the complex real-world situation, which may aid the understanding of the price formation process and could lead to further research questions.

[53] Smith (1985), p. 270.

to firms. The gamblers lose money voluntarily, the economy benefits and perhaps only [the expected utility hypothesis] suffers as a predictive theory for some types of agents. But the existence of such agents in futures, stock, and option markets will cause such markets to appear to be irrational by our definitions, whereas actually these markets may be performing with high allocative efficiency, given the environment, by taking wealth away from the gamblers and giving it to the hedgers, investors, and rational expectationists. Isn't Las Vegas an exchange market between gamblers (customers) and rational expectationists (casinos)? The question may be not 'Why are certain markets inefficient?,' but 'What is wrong with our interpretation of markets?' [...] I suspect that Adam Smith would wonder why there is so much modern professional interest in the internal efficiency or 'perfection' of particular markets, and so little interest in what determines the extent of markets, and how this in turn may create social gains that are more important and significant than the 'imperfections' in particular markets that are suggested by our theory of 'rational' preferences."[54]

The presence of a majority of traders who could be designated "naïve" in the experiments led to naïve option quotes and transactions in the option market, which closely followed contemporaneous prices in the stock market (an example of such a pair of price paths is shown in Fig. 12 below, and similarly detailed figures for all rounds of all sessions are provided in Figs. A.1–A.7 in Sect. 6.2 of the appendix). This close link between irrationally high contemporary stock and option prices could be observed in most experimental sessions, which answers the question regarding the option market's helpfulness from above: The option market *could not* be informative for subjects' expectation formation, because its prices were equally biased as those of the stock market.[55] Table 14 lists a number of statistics designed to underline this observation. If the experimental market had been efficient, it would have exhibited a number of characteristics with regard to the interrelation of the time series of stock and option prices. First, the mean stock price per period should have correlated perfectly positively with the (deterministic and public-information) fundamental value of the stock. Second, the mean option (strike) price per period in the DO8 treatment should not have been correlated with the stock price, since the rational option price in this treatment has a variance of zero, as it is fixed at 192 cent. In the DO5/10/15 treatment, the case is not as obvious. Options with a maturity at the end of period 5 have a rational (strike) price of 264, those with maturity in period 10 have a rational strike price of 144 and those maturing in period 15 have a rational strike price of 24. The only clear statement

[54] Smith (1985), footnote 7, pp. 270–271.

[55] Part of the motivation for the digital option treatments was the reported success of the Porter and Smith (1995) futures treatment. Out of a total of six bubble measure observations reported in their article (the three bubble measures amplitude, duration and turnover, calculated for two levels of experience), four showed an improvement and two a deterioration when comparing the futures treatment to the baseline setting. In light of the results presented in this study, it would be an interesting topic for future studies to investigate whether this overall improvement is a spurious result or if not, what causes underlie the different impacts of futures and digital option markets on experimental spot market efficiency.

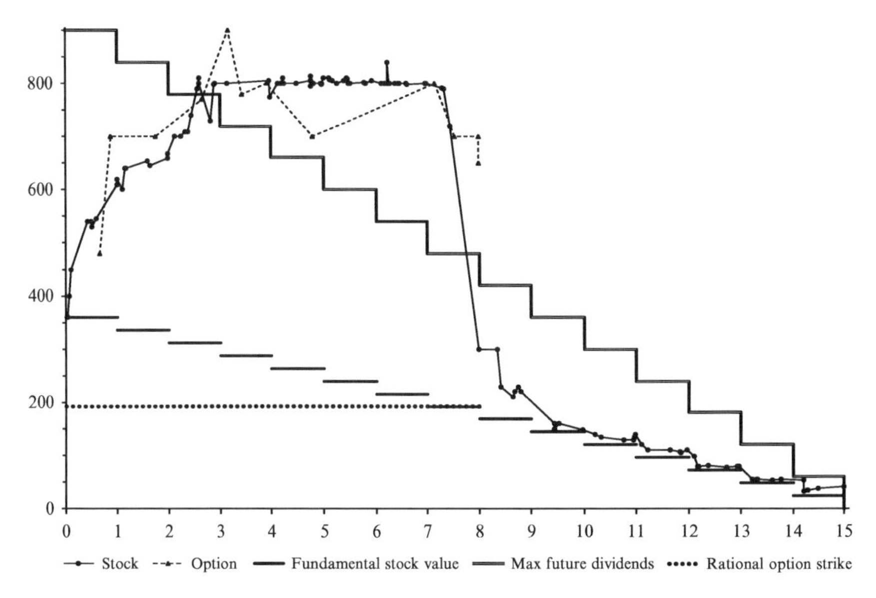

Fig. 12 Example of the price plots of an experimental session. The figure plots the stock price in cents (solid line with circles), option price (broken line with triangles), fundamental stock value (stepwise decreasing function, single solid line), rational option strike price (dotted line) and maximum value of future dividends (stepwise decreasing function, double solid line) in the second round of session 4 (DO8). The structural break in period 8 – from prices increasing relative to the fundamental value, to prices closely tracking the fundamental value – is clearly visible

regarding the mean period standard deviation of option prices in this treatment is that it is bounded from above by 120.[56]

The results reported in Table 14 depart widely from this efficient yardstick. The first content row lists the Pearson correlation between mean period stock and option prices, for all rounds of the two treatments DO8 and DO5/10/15. The stock and option prices are highly positively correlated (Pearson correlation between 0.662

[56]Since the distribution of option transactions between different periods within one round is indeterminate, the variance of an option market populated only with rational traders is also indeterminate. Its upper bound can be arrived at by exploring the transaction pattern that, while still informationally efficient, maximizes the observed option price standard deviation. This pattern is characterized by an equal number of option transactions in the intervals of periods 1–5 and periods 11–15, with no transactions in periods 6–10. Assuming that all transactions take place at the efficient prices of 264 and 24 in periods 1–5 and 11–15, respectively, the resulting (rational) standard deviation would be 120.
Note that if one carries the concept of rationality one step further, a no-trade theorem argument would suggest that there should be no transactions at any price. See Sect. 2.4.1.5 for more on no-trade theorems and their applicability to the current setting.

Table 14 Option market efficiency statistics

Measure	DO8				DO5/10/15			Both			
Round:	1	2	3	All	1	2	All	1	2	3	All
Correlation Stock – Option	0.811	0.662	0.677	0.730	0.943	0.885	0.914	0.868	0.757	0.677	0.803
Correlation Stock – Fundamental	0.309	0.857	0.927	0.621	−0.040	0.751	0.355	0.160	0.811	0.927	0.515
Correlation Option – Fundamental	−0.914	−0.512	−0.463	−0.685	−0.100	0.768	0.334	−0.565	0.037	−0.463	−0.278
Standard deviation of option prices	73.3	56.2	19.5	59.7	135.2	167.7	151.4	99.8	104.0	19.5	96.4

This table lists the mean correlation of mean period prices by round, over all sessions, by treatment. Also included is the standard deviation of option prices and 0.943[57]). An even stronger sign of the inefficiency of the option market is provided by the option standard deviation results. The mean option price standard deviation in the DO8 (DO5/10/15) treatment – which should be zero (smaller than 120) – varies between 19.5 and 73.3 (135.2 and 167.7), an interval that does not even contain the efficient figures.[58]

Table 15 gives another indication of the inefficiency of the option prices. In a market populated only with rational, risk-neutral decision-makers, the mean option price would equal the mean stock price only in period 8 (periods 5, 10, and 15) of the DO8 (DO5/10/15) treatment.[59] In all other periods with option transactions, the mean stock price would be strictly higher than the mean option strike price, since the efficient option strike price equals a future stock price, and the fundamental value of a share of stock declines deterministically and monotonously over time (measured in periods). Table 15 lists the p-values from a non-parametric, one-sided Mann-Whitney-U-Test of the null hypothesis of the mean option strike price having been larger than the mean stock transaction price (upper numbers, not in parentheses). In the DO8 (DO5/10/15) treatment, the p-values should be close to 0.5 in period 8 (5, 10, and 15), and close to 0 in periods 1 to 7 (1–4, 6–9, and 11–14). In line with theory, the null hypothesis of the mean option strike price having been larger than the mean stock transaction price could not be rejected in the option maturity periods, with some p-values close to 0.5. However, it was also not rejected in the non-option-maturity periods, contradicting the prediction of economic theory that it should have been rejected in all but period 8 (5, 10, and 15) in the DO8 (DO5/10/15) treatment.[60]

The numbers in parentheses in Table 15 are the p-values from a two-sided Mann-Whitney test of equal means, applied to the transaction prices in the stock and option markets. In the DO8 treatment, the p-values in brackets should have been close to 1 in period 8 and close to 0 in periods 1–7. This hypothesis held up relatively well in the DO8 treatment, where the pattern emerged as described when analyzing transactions in the same period over all rounds. An exception was period 8 of the first round, where the rejection of the null hypothesis of equal means

[57]Note that the Pearson correlation for the DO5/10/15 treatment is actually remarkably close to the efficient mark. While naturally not an optimal measure for the analysis of the relationship between the two variables in this case, the Pearson correlation would be 0.945 in an efficient market.

[58]The standard deviation reported in Table 14 was calculated as the mean over all sessions of the standard deviation of all transaction prices within each round. Table A.1 in Sect. 6.3 in the appendix contains additional correlation figures measuring interrelationships between a number of variables describing the experiments.

[59]These periods, which contain (at their end) an option maturity date, are shaded gray in Table 15. In the DO8 treatment, the option market is closed from periods 9 through 15 and the stock price cannot be compared to an option price, which is why there are blank spaces in Table 15 for these periods in treatment DO8.

[60]Note that, due to the fact that the table contains 23 observations from the DO8 treatment and 30 observations from the DO5/10/15 treatment (excluding the p-values over *all* rounds per treatment), even assuming purely random data, the table should be expected to display 1 value significant at the 99%-level, 3 values significant at the 95%-level and 5 values significant at the 90%-level.

Table 15 Test of differences in mean transaction prices in stock and option markets

	DO8				DO5/10/15		
Round:	1	2	3	All	1	2	All
Period	(n=4)	(n=4)	(n=1)		(n=3)	(n=3)	
1	0.326	0.272	0.525	0.390	0.457	0.229	0.417
	(0.0186[b])	(0.0023[a])	(0.9135)	(0.0280[b])	(0.5485)	(0.0041[a])	(0.1431)
2	0.279	0.212	0.200	0.355	0.527	0.299	0.450
	(0.0020[a])	(0.0016[a])	(0.0595[c])	(0.0045[a])	(0.7182)	(0.0466[b])	(0.4091)
3	0.000[a]	0.169	0.024[b]	0.329	0.417	0.408	0.578
	(0.0000[a])	(0.0003[a])	(0.0218[b])	(0.0048[a])	(0.4520)	(0.3449)	(0.2702)
4	0.073*	0.323	0.000[a]	0.265	0.461	0.497	0.503
	(0.0000[a])	(0.0680[c])	(0.0491[b])	(0.0000[a])	(0.7283)	(0.9805)	(0.9702)
5	0.234	0.096[c]	0.146	0.226	0.530	0.492	0.550
	(0.0021[a])	(0.0001[a])	(0.0649[c])	(0.0000[a])	(0.7148)	(0.9296)	(0.4153)
6	0.139	0.254	0.333	0.336	0.444	0.326	0.422
	(0.0001[a])	(0.0090[a])	(0.4037)	(0.0087[a])	(0.5197)	(0.0310[b])	(0.1809)
7	0.439	0.203	n/a	0.402	0.551	0.438	0.442
	(0.5688)	(0.0059[a])	(n/a)	(0.1904)	(0.5972)	(0.5217)	(0.3821)
8	0.258	0.581	0.464	0.429	0.451	0.578	0.453
	(0.0069[a])	(0.4420)	(0.9070)	(0.2793)	(0.6107)	(0.5106)	(0.5250)
9					0.701	0.582	0.671
					(0.0647[c])	(0.5240)	(0.0372[b])
10					0.891	0.194	0.665
					(0.0002[a])	(0.0068[a])	(0.0304[b])
11					0.389	0.128	0.324
					(0.1624)	(0.0000[a])	(0.0017[a])
12					0.405	0.087[c]	0.306
					(0.3480)	(0.0000[a])	(0.0047[a])
13					0.476	0.452	0.489
					(0.8319)	(0.7256)	(0.0899*)
14					0.414	0.265	0.305
					(0.4996)	(0.0400[b])	(0.0188[b])
15					0.747	0.560	0.618
					(0.0803[c])	(0.5690)	(0.1605)

[a]Significant at the 1%-level, [b]significant at the 5%-level, [c]significant at the 10%-level.
This table lists p-values from a one-sided Mann-Whitney (Wilcoxon rank-sum) non-parametric test of the null hypothesis of the mean option strike price being larger than the mean stock transaction price, by period, round and treatment. The numbers in parentheses are p-values from a two-sided Mann-Whitney test of the null hypothesis of equal means of transaction prices in the stock and option markets. The numbers in parentheses in the title row list the number of sessions providing observations for a given treatment-round tuple. Observations at option maturity dates are shaded gray

indicates that inexperienced subjects did not trade stocks and options at the same price. With increasing experience over rounds two (p-value of 0.44) and three (p-value of 0.91), however, the probability of the observations stemming from populations with equal means approached unity.

The picture for the DO5/10/15 treatment was less positive. There were very few cases where stock and option prices were significantly different, with the possible exception of period ten, where they were, but should not have been. Apart from the default explanation of observing spurious results, a possible reason for this curious finding could be the tendency of subjects to "manipulate" the stock price in option

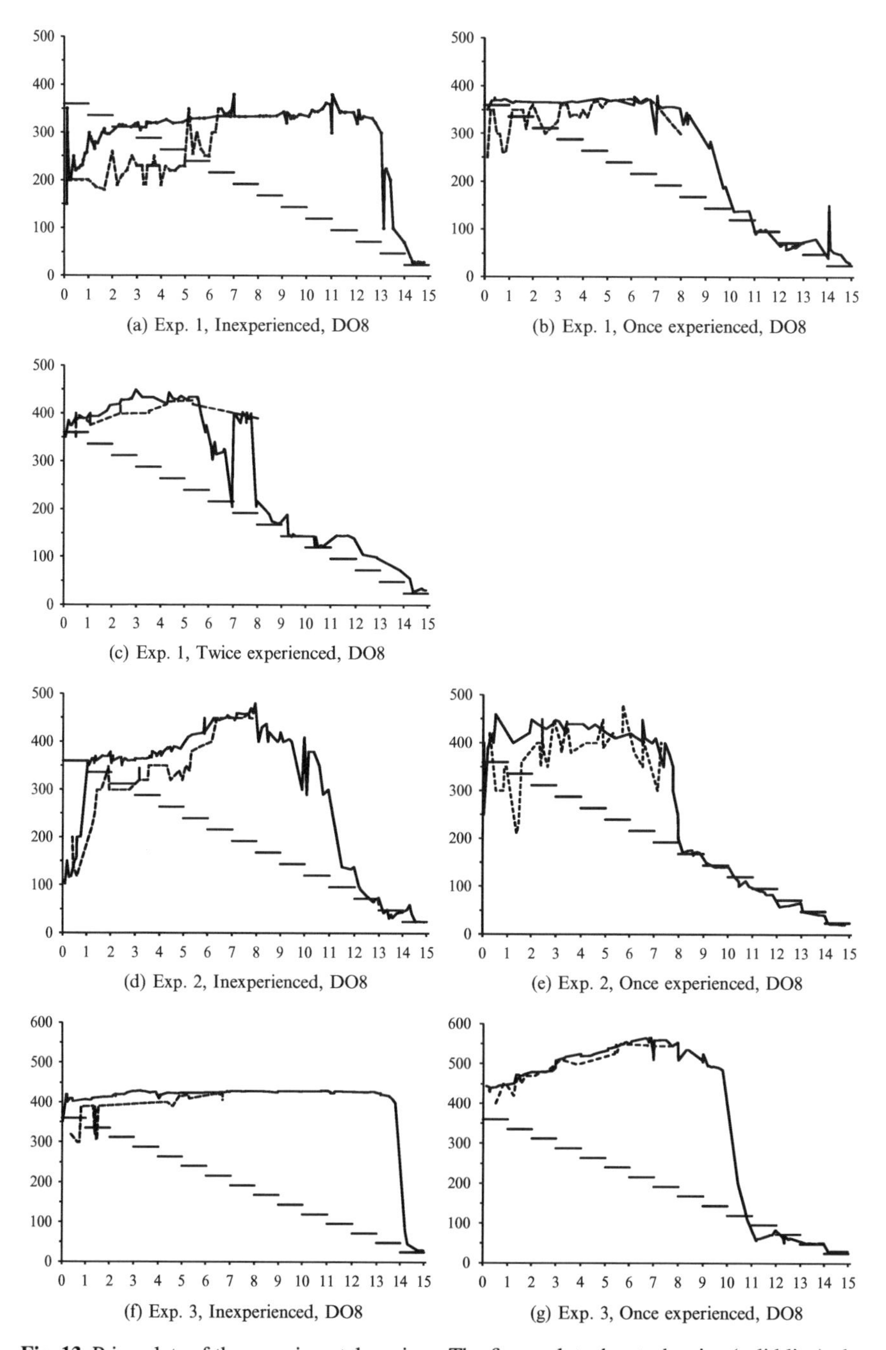

Fig. 13 Price plots of the experimental sessions. The figure plots the stock price (solid line), the option price (broken line), and the fundamental value (stepwise decreasing function) in all rounds of the experimental schedule

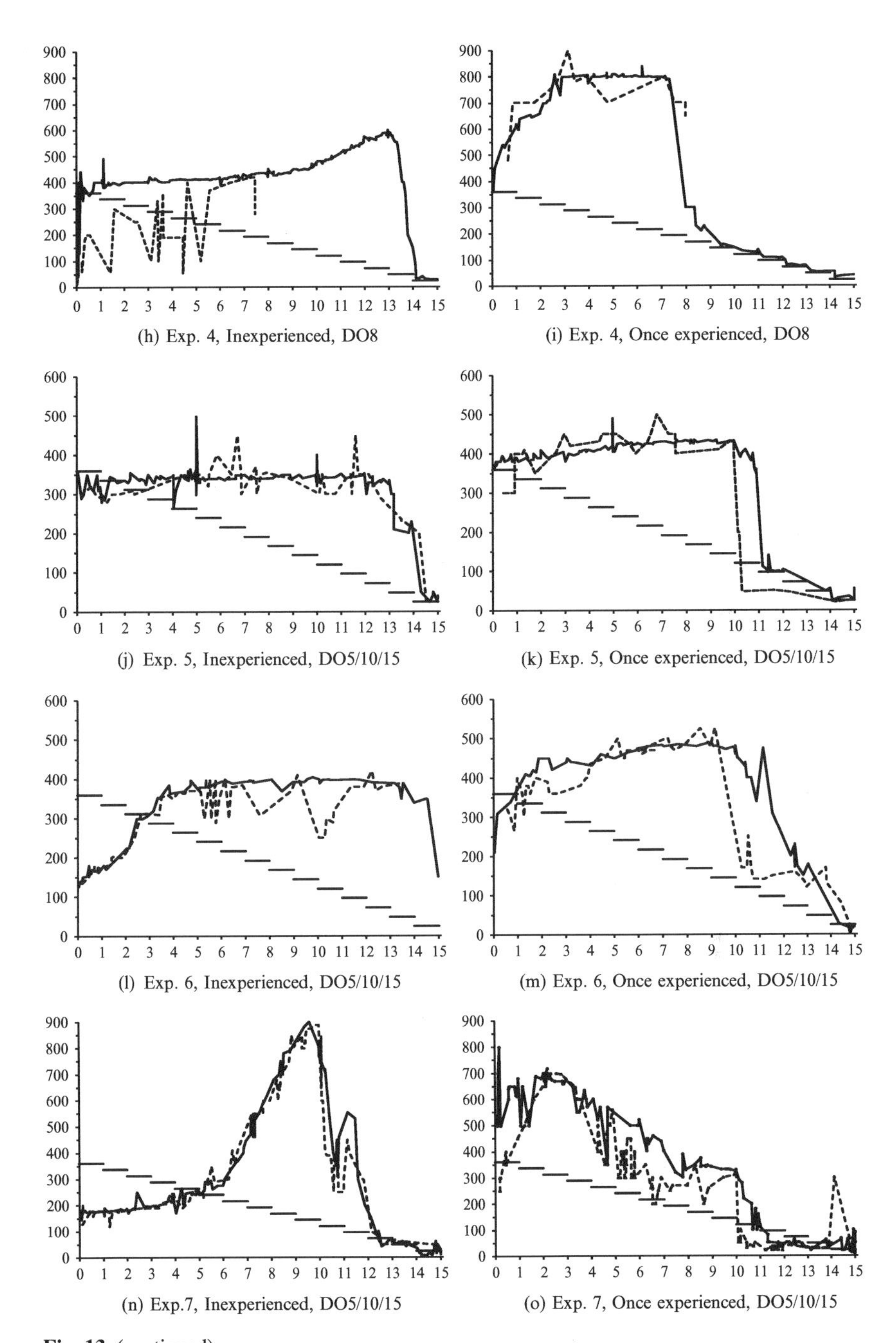

Fig. 13 (continued)

maturity periods in order to secure payoffs from their option positions. Nonetheless, this argument does not explain the lack of similar findings in periods 5 and 15, which are subject to the same trader behavior.

Another manifestation of the bounded rationality of traders in this type of market is the structural break between the eighth and ninth period in the DO8 treatment. This break is especially apparent in the sessions with experienced traders, where it is present in three of the four once-experienced treatments (see Fig. 13 above). Note also that the decline in stock prices began no sooner than the end of period eight in all of the sessions belonging to the first four experiments (with the third session in the first experiment arguably being an exception). Observing this, the author gained the impression that the option maturity date at the end of period 8 was accompanied by a break in subjects' expectations formation regime. As stated before, it seemed that during the first half of the experiment, where options on the price of the asset at the end of period 8 could be traded, the attention of subjects with regard to the stock price was focused on the levels of the strike prices of the options they held. In an attempt to further investigate this phenomenon, the DO5/10/15 treatment was designed in the hope that the more frequent option maturity dates would cause an earlier return of prices to fundamental values. Unfortunately, while this alternative treatment removed the tendency of the structural change to take place between the eighth and ninth period, it did not improve overall price efficiency.

Chapter 5
Conclusion and Outlook

If all economists were laid end to end, they would not reach a conclusion.

George Bernard Shaw 1856–1950

5.1 Summary of the Contribution

This study explored the causes and properties of price bubbles in financial asset markets. Such price bubbles have played an increasingly important role in the recent economic literature, in line with an increasing awareness among economists of their large impact on informational and allocational market efficiency. The verdict in previous studies was mixed, with diverging opinions on the existence of bubbles in recent history, on their causes and necessary conditions for their occurrence, and on their impact on financial markets. Since the seminal work of Smith et al. (1988), bubbles have also been the focus of studies employing the experimental method. Virtually hundreds of experimental runs have been conducted, varying variables like the subject pool, the dividend structure, market mechanics, the information set, monetary incentives, and the number and kind of markets operating sequentially or simultaneously. While most such experimental tests failed to significantly reduce the bubble phenomenon, factors like the frequency of dividend distributions, the fundamental value process, and most importantly subject experience have been identified as having the potential to reduce or even reverse the direction of observed bubbles.

The method employed to explore the impact of adding a digital option market to a setting similar to those employed in Smith et al. (1988) was that of experimentation. The rationale behind this approach was the increasing number of, and trading volume on, online prediction markets. These markets give investors a trading outlet which lets them bet on the future prices of various stocks and commodities. This trading activity aggregates diverse information from the relatively heterogeneous market

S. Palan, *Bubbles and Crashes in Experimental Asset Markets*,
Lecture Notes in Economics and Mathematical Systems 626,
DOI: 10.1007/978-3-642-02147-3_5, © Springer-Verlag Berlin Heidelberg 2009

participants and thus possibly influences the underlying financial markets. The specific hypothesis explored was that trading in the option market would induce subjects to form expectations about future prices at an early point in time. Having formed these expectations, and having communicated them to all participants through the public-information datum of the option price, it was then conjectured that subjects would use these expectations to derive a spot price expectation closer to fundamental value by inducing backward from the expected future prices revealed by the option market. Through this process of "looking into the future" through the lens of the option market, price bubbles would theoretically be nipped in the bud. Seven experimental sessions with a total of 86 individuals trading for more than 500 h were used to empirically test this hypothetical process. All sessions were run in university computer labs, using the z-tree software of Fischbacher (2007) and a program written and tested entirely by the author. The results were collected, preprocessed in Microsoft Excel 2007, and finally analyzed analytically using (mainly) Stata 10. Questionnaire answers collected from the subjects after each round were used to interpret the findings and to support the argument in this text.

Unfortunately, the primary hypothesis received no support from the experimental results. The extent and form of the stock price bubbles observed in the experimental markets were appraised with a variety of measures used in previous articles. The results were generally found to be comparable to or worse than those of earlier baseline experiments documented in the literature. This finding, together with previous findings of Lei et al. (2001); Haruvy et al. (2007) and others, point to one underlying cause for the inefficient price patterns observed in Smith et al. (1988)-type markets: The bounded rationality of traders. Lei et al. (2001) proved that subjects in this kind of market sometimes act irrationally, while Haruvy et al. (2007) provided evidence that subjects employ an adaptive, boundedly-rational learning rule. In this book, the Hypothesis of Observational Belief-Adaptation was formulated to describe the empirically observed learning behavior of the experimental subjects, which can succinctly be described as a rule of learning-by-observation. In a Smith et al. (1988)-type market, the application of such a rule leads to a feedback loop in which the learning of subjects in one round influences their actions in the next, which in turn influences market prices in this new round and thus provides new opportunity for learning. Through this feedback loop, and due to the fact that profitable price patterns in financial markets tend to self-destruct, the experimental markets over time converge to efficient prices. This convergence process closely resembles the process that would ensue if subjects were getting progressively more rational. This latter process of subjects becoming more rational has been the base hypothesis of most previous studies, yet has been found to be inadequate for explaining all observed facts. Conversely, the Hypothesis of Observational Belief-Adaptation – suggested by the Haruvy et al. (2007) results and formulated in this book – is a new theory to be explored in the quest to explain the bubble-and-crash pattern in Smith et al. (1988)-type markets. As Vernon Smith himself put it so eloquently:[1] "Well-formulated theories in most

[1] Smith (1994), p. 114.

sciences tend to be preceded by much observation, which in turn stimulates curiosity as to what accounts for the documented regularities". In this vein, the Hypothesis of Observational Belief-Adaptation is offered as an observation that might help scholars in deriving future research questions and exploring the causes of bubble-and-crash patterns in financial market prices.

As an additional contribution, this book offers a standardized compilation of bubble measures for Smith et al. (1988)-type markets. This listing contains not only an overview of the varied measures employed in the pertinent literature, but also lists and calculates more than 450 results for these measures in a variety of different treatment designs. These measure results are provided in Tables 7 through 11. Their first role was to put the findings reported for the current experiment into perspective with those from previous experiments from the literature. A process of benchmarking by comparing the results then led to the conclusion that the primary research hypothesis could be rejected. The bubble measures also serve a second role, however, which – given the experimental findings – is likely to prove more important for the academic discipline. This second role is as a reference for the comparison of the outcomes of future work with the established results in this field. By adhering to standardized metrics for the evaluation of a treatment's effects, future studies will be more readily comparable, and significant factors for the explanation of empirical results will be easier to isolate. The choice of the metric used for the presentation of a study's outcomes may have considerable impact on the interpretation of its results. For that reason, a standardized presentation of experimental results is preferable to choosing the reported measures in a discretionary way. It precludes one factor of variation that might otherwise lead to drawing erroneous conclusions which are not caused by differences in the experimental outcomes, but only in their presentation.

5.2 Outlook and Suggestions for Future Research

How is it that the pricing system accomplishes the world's work without anyone being in charge? Like language, no one invented it. None of us could have invented it, and its operation depends in no way on anyone's comprehension or understanding of it. Somehow, it is a product of culture; yet in important ways, the pricing system is what makes culture possible. Smash it in the command economy and it rises as a Phoenix with a thousand heads, as the command system becomes shot through with bribery, favors, barter and underground exchange. Indeed, these latter elements may prevent the command system from collapsing. [...] The pricing system – How is order produced from freedom of choice? – is a scientific mystery as deep, fundamental, and inspiring as that of the expanding universe or the forces that bind matter. [...] But what can we as economists say for sure about what we know of the pricing system? It would appear that after 200 years, we know and understand very little.

Vernon L. Smith (1982)

Past studies often conjectured that the inefficiency of prices in experimental asset markets was due to speculation. Such speculation can arise because subjects face uncertainty regarding the behavior of their fellow subjects. The *actual presence* of irrational traders is not a necessary condition for bubbles, goes the argument. The *possibility* of their existence is a sufficient condition to cause rational traders to trade at prices deviating from fundamental values. This is true because they can both expect to reap capital gains and to reap dividend income by transacting at a later point in time, where prices might be even farther out of line with fundamentals. This point was made very clear in a number of studies, including Smith et al. (1998); Porter and Smith (1995); and Plott (1991). In contrast to this, Lei et al. (2001) found that irrational traders are present in this type of experiment, since they are the only possible explanation for the bubbles they observed in their no-speculation treatment. It is not merely the possibility of the presence of irrational traders that causes bubble-and-crash patterns in groups of fully rational subjects, but rather their actual participation.

One interesting research question is what factor is responsible for subjects starting to act irrationally in the context of the studies discussed in this text, while they act very rationally in other experimental studies in the field of financial economics. Since the double auction mechanism works at remarkably high efficiency levels when employed in conjunction with externally induced demand and supply schedules that differ among subjects (cp. e.g., Smith (1962)), it can be dismissed as the sole cause of inefficiency. As Lei et al. (2001) showed, the possibility of traders to act in both the role of buyer and seller is also no necessary condition for irrational behavior. Porter and Smith (1995) ruled out that the risk posed by uncertain dividends causes investor irrationality. –Before running down the list of possible factors (and combinations thereof), this discussion section will limit itself to pointing out this issue as a promising future research venue and offering one possibly promising observation: A pattern in the observed irrationality of subjects in the context of all Smith et al. (1988)-type markets is that they seem to place an unfoundedly strong weight on dividends, an observation that will be referred to as hypothesis H3 below. In the usual case of a positive expected dividend payment at the end of each period, subjects overvalue the stock, while in the case of a negative expected dividend (see Davies (2006)), they undervalue it. If there are no repeated dividend payments, as in Smith et al. (2000), the bubble disappears completely (in nine out of ten sessions), and in treatments with fewer dividend payouts it is smaller than in markets with more dividend payouts (cp. Ackert et al. (2006c)). In a similar vein, Noussair and Tucker (2006) wrote:[2]

> "Lei et al. (2001) argue that in addition to speculation, decision errors on the part of market participants also play a role in bubble formation. These errors appear to originate in an inability on the part of traders to correctly value the asset by linking the expected future dividend stream to a rational limit price, as well as in the procedures of the experiment, which encourage active participation in the market due to a lack of alternative activities.

[2]Noussair and Tucker (2006), p. 168.

> These effects, both speculation and decision error, appear to us to provide the most reasonable account of the source of the bubble and crash phenomenon".

Considering this evidence, it might prove worthwhile to test whether a similar effect prevails in real-world markets. Given the extensive data on financial markets available to financial economists, running a regression of excess returns on dividend yields for stocks (or on a dummy variable identifying (non-)dividend-paying stocks) during a market bubble like that in October 1987 should prove relatively straightforward. While dividends have been associated with excess returns for a number of years,[3] no study that the author is aware of investigated their role in bubble markets. A test like the one described above would reveal whether the excess attention paid to dividends in the experimental context and its impact on bubble formation is an artifact of the laboratory or can be considered a general feature of financial markets.

Another interesting question to explore would be the result of an experimental institution where all option markets in the DO5/10/15 treatment are opened simultaneously, to give subjects an outlet to trade on their expectations of the stock price far in the future. As stated in Sect. 2.4.4.2, Noussair and Tucker (2006) suggested a similar treatment after reporting on their success in eliminating bubbles by installing a futures market for every period, opening in reverse order of their maturity dates. They suggested that such a design helps traders to apply backward induction in their expectations formation. The difference between the treatment suggested here and the design employed in Noussair and Tucker (2006) is that the former is more realistic and could be employed in financial markets. Since stocks in financial markets usually do not have a pre-determined liquidation date, opening futures markets in reverse order (i.e., backwards through time, starting from the liquidation date), as in Noussair and Tucker (2006), is impossible. On the contrary, opening new or additional, low-cost futures or option markets on stocks is unproblematic and in fact in the process of being implemented by the existing online prediction markets.

Finally, future research is required to more thoroughly investigate the Hypothesis of Observational Belief-Adaptation. It was created only after a number of experiments had been run and therefore had to be very loosely formulated here. The Hypothesis of Observational Belief-Adaptation requires operationalization and testing to judge its validity in describing subjects' behavior in experimental asset markets. Likewise, the existence of two distinct sets of traders, one naïve and one rational, became apparent only after a number of sessions had been conducted and the quantitative results and questionnaire answers had been analyzed. This precluded focusing on these phenomena from the beginning. For this reason, the scope of analyses exploring them based on the data collected for this book was limited. Future research should aim at solidly establishing the validity of the Hypothesis of Observational Belief-Adaptation and the existence of these two groups of subjects, using both experimental and classical empirical methods.

[3] Cp. e.g., Siegel (2005).

Once (and if) their existence has been firmly established, characteristics of markets with differing ratios of naïve to rational traders could be made the focus of research. This could in turn yield insights into the functioning of markets with, for example, differing populations of professional and individual investors. It would also be interesting to see whether these two groups correspond to the two groups of naïve and rational traders observed in the experiments for this study, or whether the distinction between professional and individual investors is not as clear-cut.

To summarize, this book proposes the following new research hypotheses, which could not be probed with the existing data set:

H1: Common expectations and convergence to fundamental value in Smith et al. (1988)-type experiments are due to learning from observation, not due to logic applied to common information. This conjecture shall be referred to as the *Hypothesis of Observational Belief-Adaptation.*

H2: Traders can be assigned to the two groups of (1) naïve traders and (2) rational traders. They differ with regard to their grasp of the market mechanics, of the stock's fundamental value process, and with respect to their expectations formation.

- *H2i:* Naïve traders follow the Hypothesis of Observational Belief-Adaptation. They trade based on prices in the current and in past rounds, but do not condition on the fundamental value.
- *H2ii:* Rational traders understand the market mechanics and the fundamental value process. In the presence of naïve traders, they initially trade based on rational expectations, but soon speculate in the expectation of being able to invert their transactions in the future, reaping capital gains or dividend income in the intervening time.

H3: Naïve traders overweight dividend payments.

- *H3i:* Experimental assets without dividend payments trade at prices close to their fundamental value.
- *H3ii:* For experimental assets with positive expected dividend payments, increases in the frequency of dividend payments increase the number of transactions at prices exceeding the fundamental value.
- *H3iii:* For experimental assets with negative expected dividend payments, increases in the frequency of dividend payments increase the number of transactions at prices below the fundamental value.

Chapter 6
Appendices

6.1 Explanation of Bubble Measure Calculations

This section explains the origin of the bubble measures reported in Tables 7 through 11 in Sect. 4.1.1.

Ackert et al. (2006)

The measures reported for Ackert et al. (2006) were gathered as follows: Only data on the standard (non-lottery) asset is reported, since the Ackert et al. results can be interpreted as outcomes regarding the impact of adding a lottery asset to a standard asset experimental market. The PositiveDurationACCD measure was taken from Table 2 on p. 428, Panel A, row 4. The information on UnderpricedTransactions and OverpricedTransactions was taken from the text on pp. 428, 430, and 431. Finally, information on turnover stemmed from Ackert et al. (2002), Table 3, p. 30, Panel A, row 7.[1]

Ackert and Church (2001)

The bubble measures were found in Table 3, p. 17, columns 3 (DurationPS), 4 (AmplitudeK), 5 (ExtremeOverpricingAC), and 7 (TurnoverK).

Caginalp et al. (1998)

The mean price time series were found in Table 1, p. 758. AmplitudeHN is equivalent to AmplitudeK, because the fundamental value was constant.

[1]Ackert et al. (2002) was a working paper preceding Ackert et al. (2006); in the latter, the turnover results were no longer reported.

S. Palan, *Bubbles and Crashes in Experimental Asset Markets*,
Lecture Notes in Economics and Mathematical Systems 626,
DOI: 10.1007/978-3-642-02147-3_6, © Springer-Verlag Berlin Heidelberg 2009

Caginalp et al. (2001)

The information on the treatments stems from the text on pages 83–84. The information on the bubble measures is taken from Tables 1a-1c, pp. 84–85, and Table 2, p. 90.

Corgnet et al. (2008)

Information on TurnoverK is taken from Table 10, p. 26 and information on AmplitudeK, DurationPS, AverageDispersion, and DeviationKSWV is taken from the descriptive statistics in Table A.1, p. 31, rows 2–4, respectively. For the *Baseline* and *Announcement low (high), preset* treatments, the latter measures were calculated from more detailed data found in the earlier version of this article, Corgnet et al. (2007). DeviationKSWV was calculated as Corgnet et al.'s (2008) "Normalized Absolute Price Deviation" measure, divided by 100. The Average-Dispersion results reported for Corgnet et al. (2008) deviate from the definition in formula (7) in that they use mean period transaction prices instead of median prices.

Davies (2006)

Since no clear results were found with regard to heterogeneous liquidity regimes, these were pooled for the two treatments of decreasing (*Baseline*) and also for increasing asset value (*Increasing value*). Due to the difference in the number of participants, the two sessions IHZ and ILZ were not included. Information on the total stock of units was taken from Table 1, p. 6. Information on trading volume and mean period prices was taken from columns 3 and 4, respectively, of the tables in Annex 4, pp. 31–35. Since no measure results (out of those used in this book) were reported, all were calculated by the author.

Dufwenberg et al. (2005)

All information was calculated from Table 1 on p. 1734. Rows 9–12 contained information on DeviationKSWV and rows 19–22 contained data on AmplitudeK. Columns 2–6 (7–9) contained data from the *2/3 experienced* (*1/3 experienced*) treatments.

Haruvy et al. (2007)

All information was taken from Table 2, p. 1908. Results for AmplitudeHN were taken from row 4, DeviationKSWV from row 5, and Turnover from row 3.

Haruvy and Noussair (2006)

All information was taken from Tables 2 and 3 on pp. 1132 and 1133. Results for AmplitudeHN were taken from column 3, and TurnoverK was taken from column 5.

Hirota and Sunder (2007)

Information about the fundamental value[2], minimum and maximum possible dividend value, and the number of periods was taken from Table 1, p. 1881, columns 4, 5, and 12, respectively. Information on the treatment designs was taken from the text and from Table 2, p. 1885. Mean period price information was taken from Figs. 1–11 on pp. 1888–1893.

Hussam et al. (2008)

All information is taken from Table 4, p. 934. The measures reported for this article are not derived by directly calculating the measure results and averaging them over all applicable rounds, but by application of a seemingly unrelated regression (SUR). See Hussam et al. (2008), p. 933 for the exact specification of the regression equations.

King (1991)

AmplitudeK is the mean of the mean price changes from p. 203, Table 2, column 4, sessions 1-6, divided by the first period fundamental value of 3, for the *Information on each period* treatment, and sessions 7–12 – also divided by 3 – for the *Information on periods 1, 6, and 11* treatments. DurationK similarly is the mean of the boom duration values from p. 203, Table 2, column 6. The same is true for TurnoverK (column 7).

King et al. (1993)

The measures were taken from Table 13.2, p. 187, Table 13.3, p. 189, and Table 13.4, p. 192. In each, column 3 contained DeviationKSWV, column 2 the sample size, column 4 the VarianceKSWV, column 5 AmplitudeK, column 6 DurationK, and column 7 TurnoverK. The Baseline series was in part taken from Smith et al. (1988), but King et al. gave no details on the exact make-up of their data in this treatment. The amplitude series was divided by 3.6, which was the intrinsic value of the share in period 1 of their experiment. This served to transform their amplitude measure into AmplitudeK, as described in Sect. 4.1.2.1. One exception was the amplitude of the *Informed insiders & short selling* treatment with inexperienced subjects, which had a first-period intrinsic value of 2.40 and was calculated accordingly. Fig. 13.3(a), p. 195 of King et al. (1993) suggests that the same is true for one of the *Limit price change rule* treatments with inexperienced subjects (304) – the second (307) was not printed, but was assumed also to have started with an intrinsic value of 2.40. The measure was adjusted accordingly.

[2]There is a typo in Table 1: The high dividend level (i.e., fundamental value) assigned in session 9 was 130, not 30, as can be seen from the 2005 working paper version of this article.

Lei et al. (2001)

TurnoverK was calculated as the mean of the three listed percentage turnover figures from Table 2, p. 841, for the *No speculation* treatment, from Table 4, p. 850, for the *Baseline* and *Two markets* treatments, and from Table 6, p. 853, for the *Two Markets & No Speculation* treatments. In all treatment designs of Lei et al. (2001), there were sessions where the dividend followed a discrete uniform probability distribution with two possible dividend values. (All *No speculation*, two of the six *Two markets*, all of the *Two markets & No speculation*, and three of the four *Baseline* treatments. The dividend values were 20 and 40 Francs.) In the mean measure results reported, no distinction was made between results from sessions using this dividend scheme and those employing the more common discrete uniform probability distribution with four possible dividend values (0, 8, 28, 60). Additionally, in one of the six *Two markets* treatments, the stock had a final buyout value. All Lei et al. (2001) treatments employed initial cash endowments, loans, etc. deviating from most other experiments and should thus be compared to these earlier results with caution.

Porter and Smith (1994)

Porter and Smith (1994) did not clearly explain their measures except for amplitude (which corresponds to AmplitudeK). For duration and turnover, the specifications of DurationPS and TurnoverK, respectively, were assumed to hold. All measures were taken from their Table 2, p. 116.

Porter and Smith (1995)

The measures stemmed from their Table 5, p. 521. AmplitudeK was reported in columns 2 and 5, DurationPS in columns 3 and 6, and TurnoverK in columns 4 and 7. The sample size was taken from Table 4 and the notes of Table 4, p. 519. In addition to the sample size listed for the Porter and Smith (1995) *Baseline* experiments in Tables 7 to 11 of this text, three *Baseline* experiments were mentioned in their article, but it was not made clear whether they were conducted with inexperienced or experienced subjects. Hence, these three experiments were included in the sample size number of neither the inexperienced nor the experienced measures. The *Switch* subjects had twice participated in certain dividend treatments and then participated in an uncertain dividend treatment. Even though they were listed in Porter and Smith (1995) as "once experienced," they are listed as "twice experienced" in Tables 7–11 of the present book.

Noussair et al. (2001)

The measures TurnoverK, AmplitudeK, and DeviationKSWV were taken from Table 2, p. 94, row 10, columns 2–4, respectively. For the *Constant value* design, AmplitudeHN equals AmplitudeK, which is why their findings are reported under both headings. The measures DurationK and DurationPS were calculated from Table B1 in their appendix, p. 101.

Noussair and Powell (2008)

All measures were calculated as averages over the five columns of the *Peak* and *Valley* treatments, respectively, of Tables A1 and A2, pp. 31–32. AverageBiasHN and AverageDispersion are calculated by dividing the mean Total Bias and Total Dispersion results by the number of periods per round (i.e., 15).

Noussair and Tucker (2006)

All measures were calculated as averages of the rows 2–5 of columns 2–4 of Table 1, p. 174.

Smith et al. (1988)

In order to calculate TurnoverK, the number of shares was determined from Table 1, p. 1126, and the turnover per period from the Figs. 2 through 13. The "x" following the session number denoted experienced subjects. For their experiment 19x (*Dividend once*, experienced) Smith et al. reported turnover only for periods 1–13. Turnover for periods 14–15 was therefore assumed to have been zero. There was conflicting information on period nine of experiment 28x in Figs. 7 and 9. Figure 7 listed the number of transactions in period nine as eight, while Fig. 9 reported it as two. The more detailed presentation in Fig. 9 suggested that the latter is the correct number, which is why it was used here instead of the former. No data was found on experiment 46f, and the data from experiments 12xn (confederate insiders), 20xpc (price control experienced), 23pc (price control inexperienced), 42xf (30 periods) was not utilized. The reason is that these treatments covered designs which were not taken up in later studies that the author is aware of. Since there is only a single observation each for these designs, they do not warrant the introduction of a new category. Another treatment variable was whether subjects had to provide forecasts of future prices. No distinction was made between treatments where this was the case and treatments where forecasts were not solicited. TurnoverK *Baseline* inexperienced contains Experiment 10, which was run with business professionals.

Smith et al. (2000)

The bubble measures were calculated as means of the numbers reported in Appendix Table 1, p. 582. DevationKSWV was taken from column 2, Variance KSWV from column 3, AmplitudeK from column 4, TurnoverK from column 5. The treatments designated "A1" referred to what is designated the *Dividend once* treatment here, "A2" corresponded to *Baseline* experiments, and "A3" was a *Dividend mix* treatment. The "x" following the session number again denoted experienced subjects. For the *Dividend once* design, AmplitudeHN equaled

AmplitudeK, which was why their findings are reported under both headings. VarianceKSWV was taken from column 3 of Appendix Table 1.

Van Boening et al. (1993)

AmplitudeK was taken from Table 1, p. 181. It was calculated by taking the mean of the price amplitude (column 2), divided by 3.75, which was the intrinsic value of the share in period 1 of their experiment. This served to transform their amplitude measure into AmplitudeK, as reported in Sect. 4.1.2.1 of this book. The same calculation (without dividing by 3.75) was performed for TurnoverK (column 4) and DeviationKSWV (calculated from the "Absolute I.V. price deviations" numbers in column 3).

6.2 Detailed Price Plots

The following Figs. A.1–7 show the stock and option transaction prices in all rounds of each of the seven sessions. Also plotted is the fundamental value of a share of stock, the maximum possible future dividend payments from one

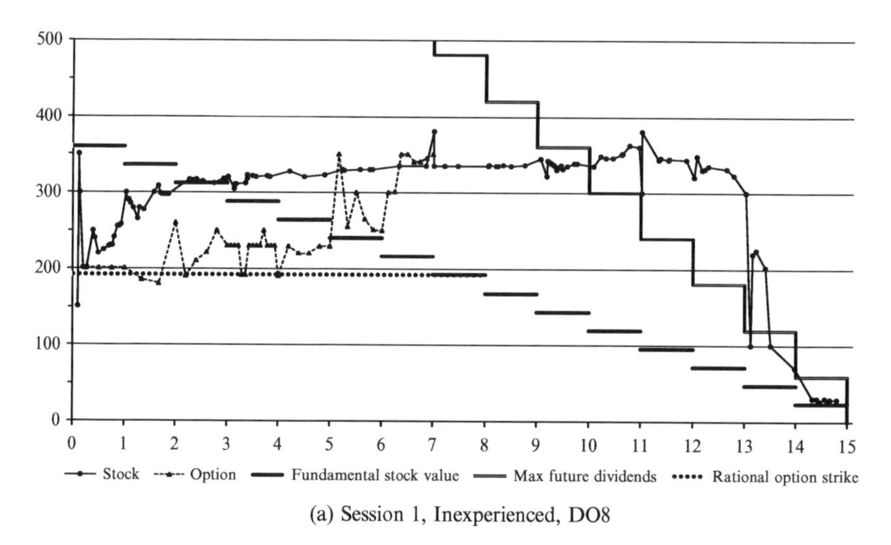

(a) Session 1, Inexperienced, DO8

Fig. A.1 Detailed Price Plots, Session 1 (DO8). The figure plots the stock price (solid line with dots), option price (broken line with triangles), fundamental stock value (stepwise decreasing function, single solid line), rational option strike price (dotted line) and maximum value of future dividends (stepwise decreasing function, double solid line) in all rounds of session 1. (a) Session 1, Inexperienced, DO8. (b) Session 2, Once experienced, DO8. (c) Session 1, Twice experienced, DO8

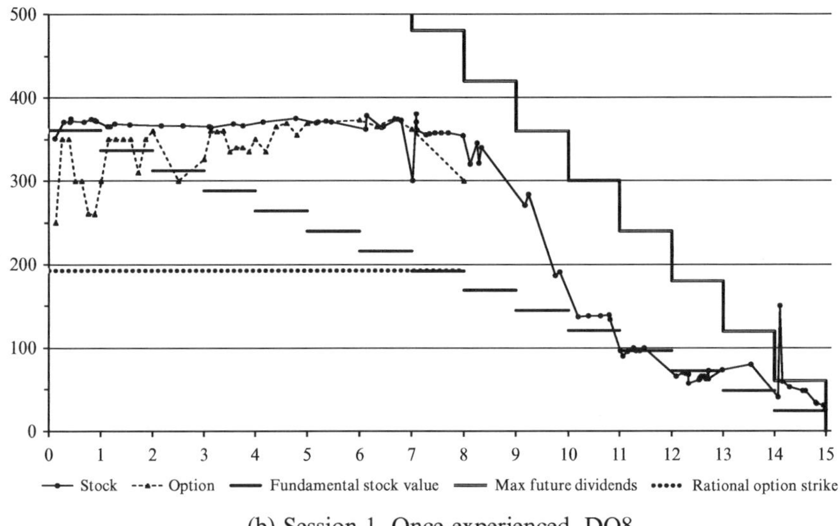

(b) Session 1, Once experienced, DO8

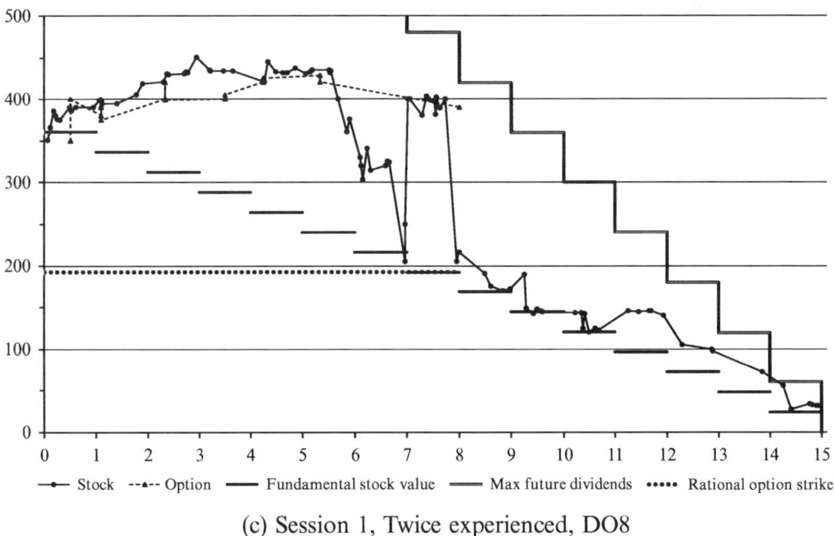

(c) Session 1, Twice experienced, DO8

Fig. A.1 (continued)

share of stock, and the fundamental share value at the time of option maturity (referred to as "Rational option strike"). Note that the scale of the vertical axis is held constant between rounds to permit easy within-session comparison, but may differ between sessions, since the maximum stock price varied strongly between sessions.

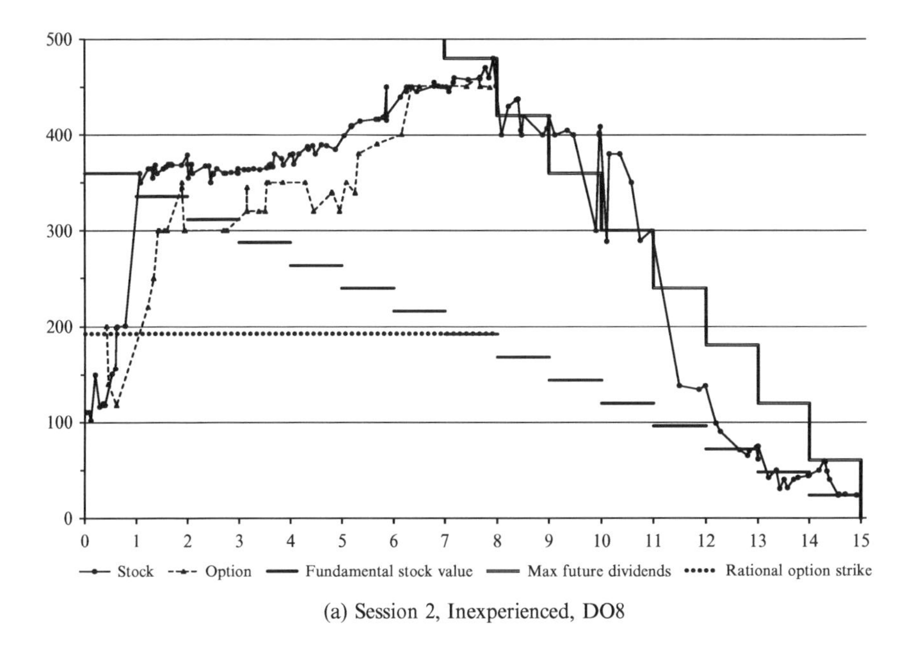

(a) Session 2, Inexperienced, DO8

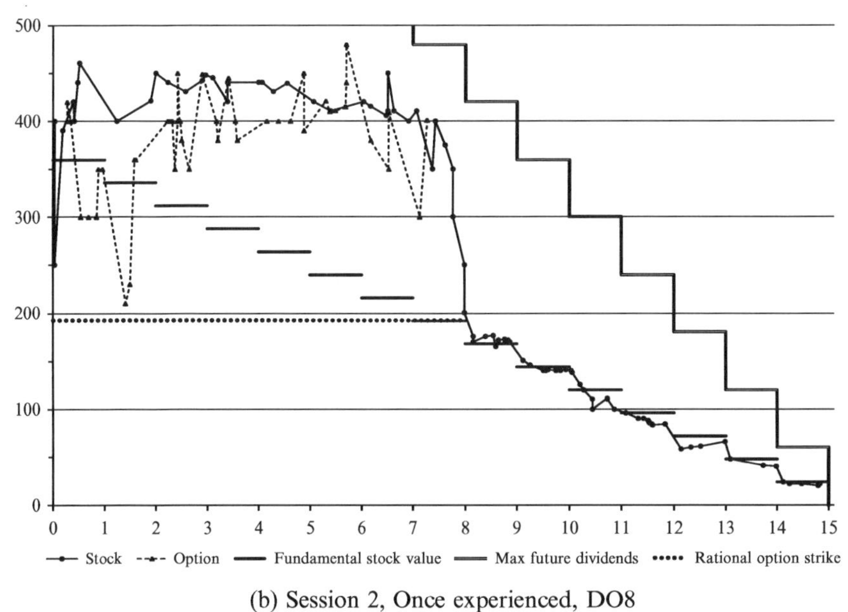

(b) Session 2, Once experienced, DO8

Fig. A.2 Detailed Price Plots, Session 2 (DO8). The figure plots the stock price (solid line with dots), option price (broken line with triangles), fundamental stock value (stepwise decreasing function, single solid line), rational option strike price (dotted line) and maximum value of future dividends (stepwise decreasing function, double solid line) in all rounds of session 2. (a) Session 3, Inexperienced, DO8. (b) Session 3, Once experienced, DO8

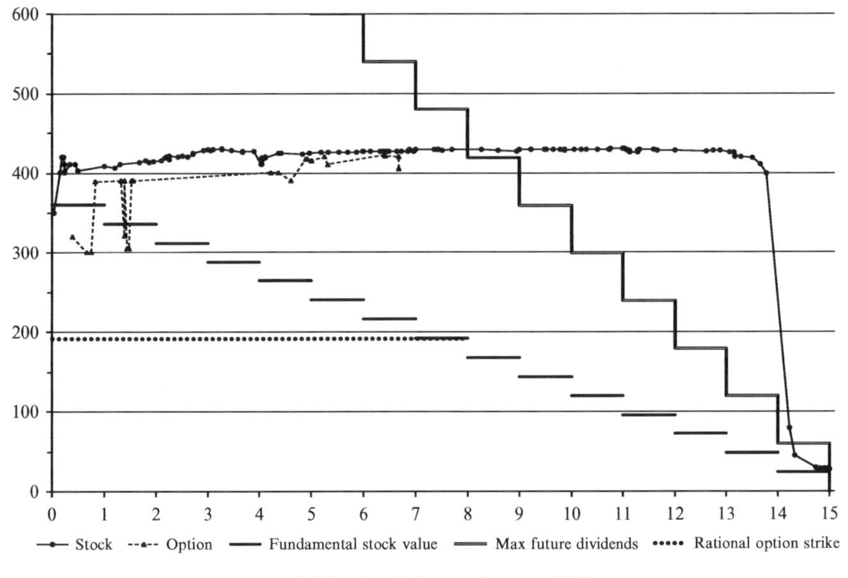

(a) Session 3, Inexperienced, DO8

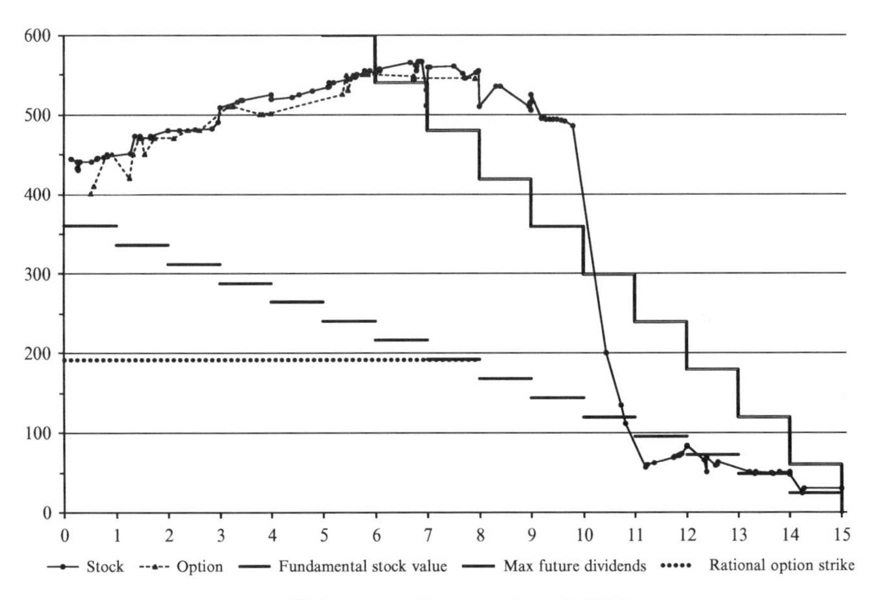

(b) Session 3, Once experienced, DO8

Fig. A.3 Detailed Price Plots, Session 3 (DO8). The figure plots the stock price (solid line with dots), option price (broken line with triangles), fundamental stock value (stepwise decreasing function, single solid line), rational option strike price (dotted line) and maximum value of future dividends (stepwise decreasing function, double solid line) in all rounds of session 3. (a) Session 4, Inexperienced, DO8. (b) Session 4, Once experienced, DO8

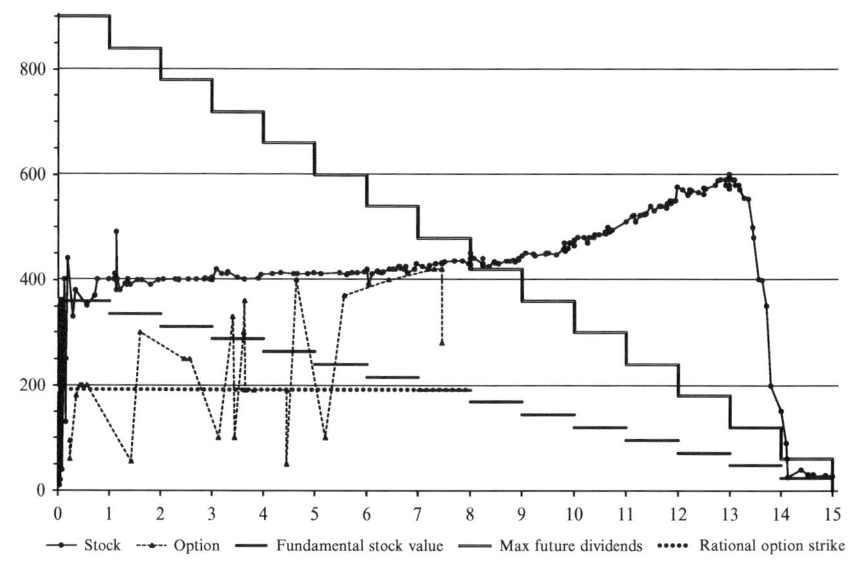

(a) Session 4, Inexperienced, DO8

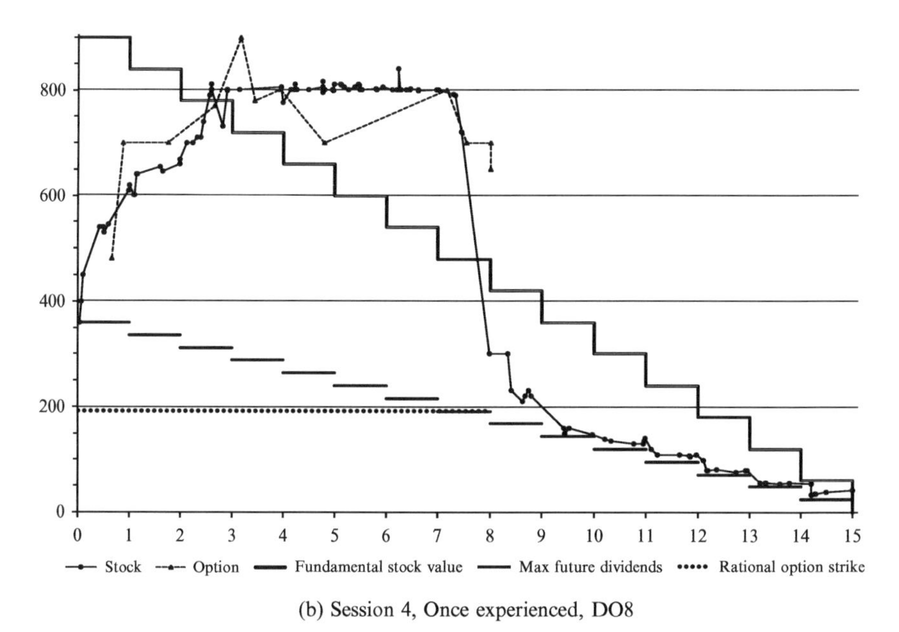

(b) Session 4, Once experienced, DO8

Fig. A.4 Detailed Price Plots, Session 4 (DO8). The figure plots the stock price (solid line with dots), option price (broken line with triangles), fundamental stock value (stepwise decreasing function, single solid line), rational option strike price (dotted line) and maximum value of future dividends (stepwise decreasing function, double solid line) in all rounds of session 4. (a) Session 5, Inexperienced, DO5/10/15 (b) Session 5, Once experienced, DO5/10/15

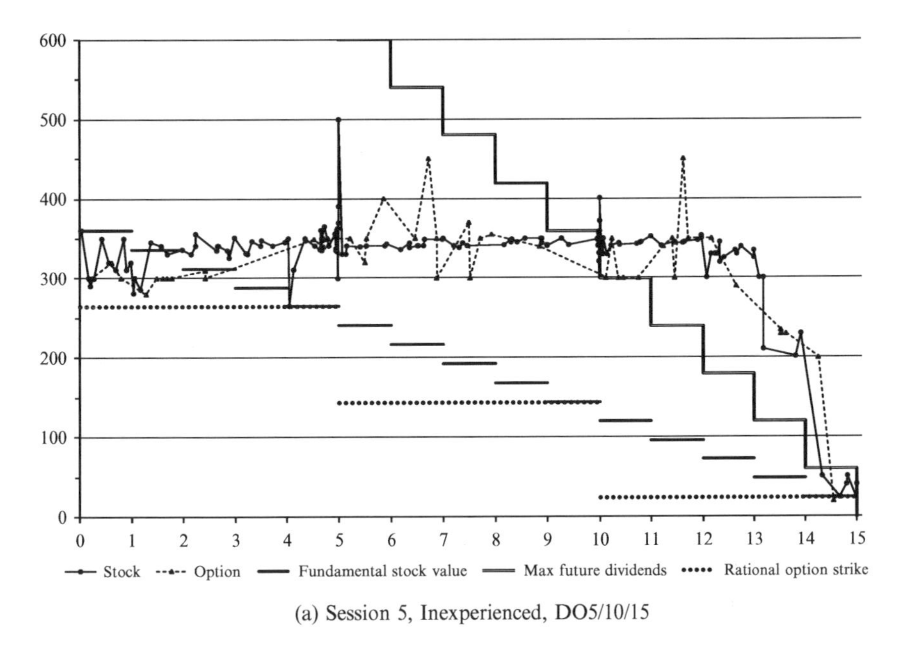

(a) Session 5, Inexperienced, DO5/10/15

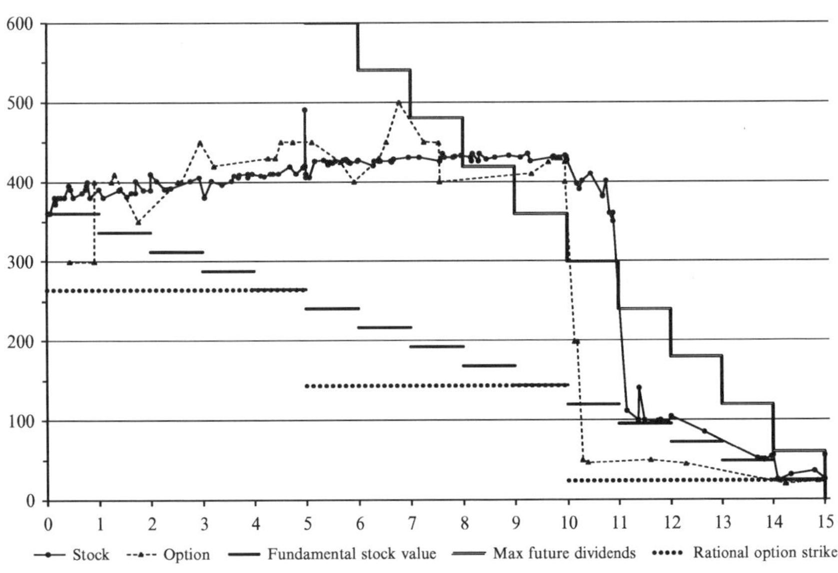

(b) Session 5, Once experienced, DO5/10/15

Fig. A.5 Detailed Price Plots, Session 5 (DO8). The figure plots the stock price (solid line with dots), option price (broken line with triangles), fundamental stock value (stepwise decreasing function, single solid line), rational option strike price (dotted line) and maximum value of future dividends (stepwise decreasing function, double solid line) in all rounds of session 5. (a) Session 6, Inexperienced, DO5/10/15. (b) Session 6, Once experienced, DO5/10/15

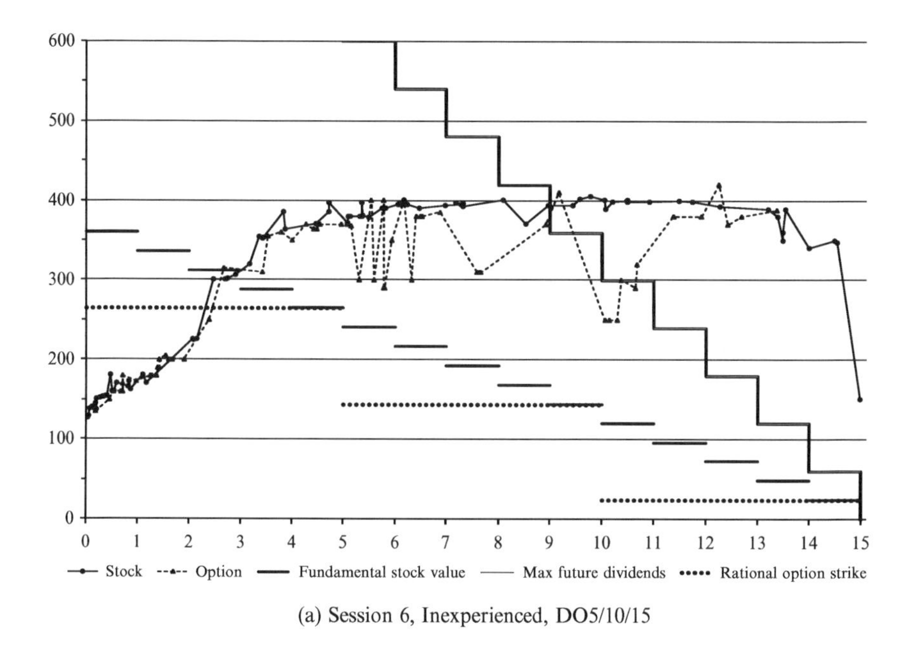

(a) Session 6, Inexperienced, DO5/10/15

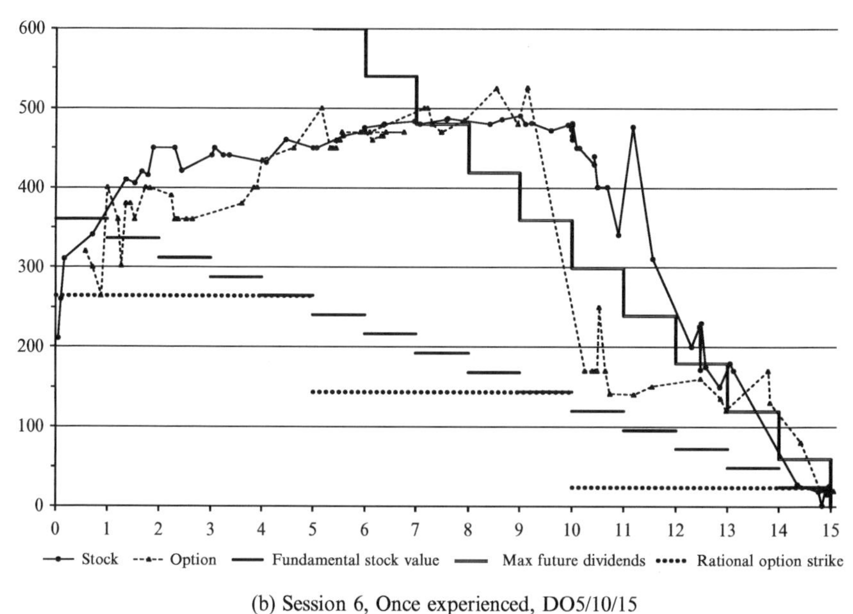

(b) Session 6, Once experienced, DO5/10/15

Fig. A.6 Detailed Price Plots, Session 6 (DO8). The figure plots the stock price (solid line with dots), option price (broken line with triangles), fundamental stock value (stepwise decreasing function, single solid line), rational option strike price (dotted line) and maximum value of future dividends (stepwise decreasing function, double solid line) in all rounds of session 6. (a) Session 7, Inexperienced, DO5/10/15. (b) Session 7, Once experienced, DO5/10/15

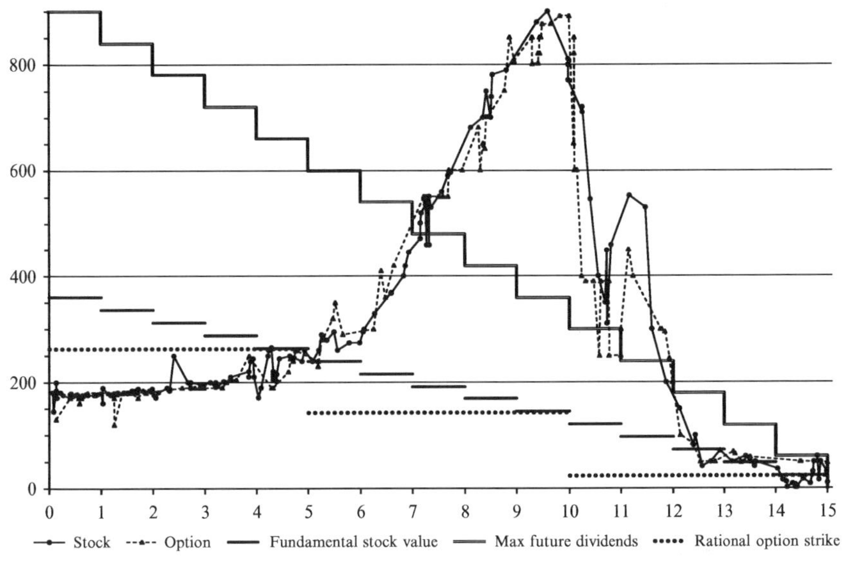

(a) Session 7, Inexperienced, DO5/10/15

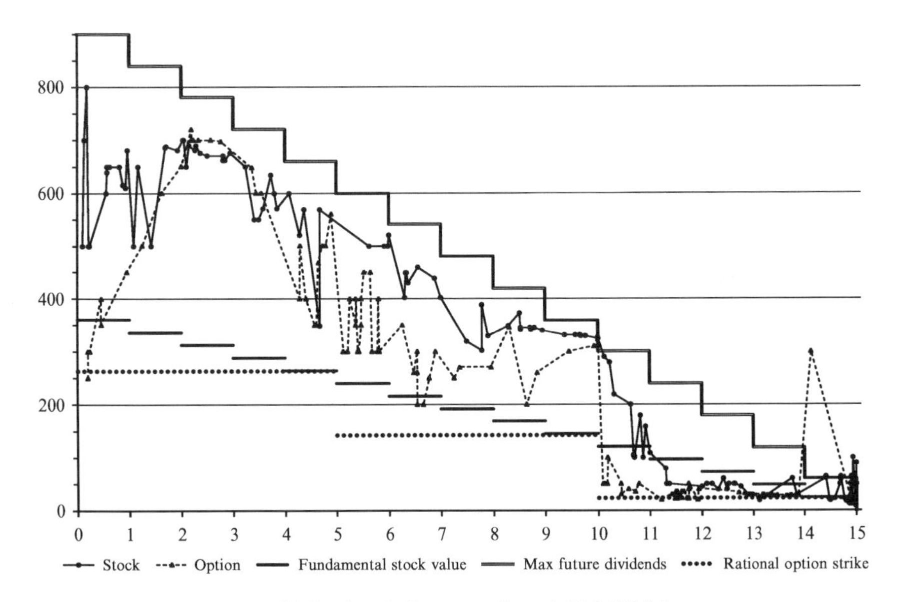

(b) Session 7, Once experienced, DO5/10/15

Fig. A.7 Detailed Price Plots, Session 7 (DO8). The figure plots the stock price (solid line with dots), option price (broken line with triangles), fundamental stock value (stepwise decreasing function, single solid line), rational option strike price (dotted line) and maximum value of future dividends (stepwise decreasing function, double solid line) in all rounds of session 7

6.3 Statistical Analysis of Questionnaire Responses

Table A.1 Correlation Matrix of Questionnaire Responses and Structural Variables

	Round	Ration	OMktHelped	Understanding	Instructions	Age	Sex	PreviousExp	FinanceKnow	TradedStock	TradedOptions	Result	StartingStock	Education	StockTrans	OptionTrans	Treatment
Round	1.00[a]																
	(0.000)																
Rational	0.04	1.00[a]															
	(0.597)	(0.000)															
OMktHelped	0.00	0.21[a]	1.00[a]														
	(0.988)	(0.004)	(0.000)														
Understanding	0.02	0.39[a]	0.18[b]	1.00[a]													
	(0.786)	(0.000)	(0.017)	(0.000)													
Instructions	0.05	0.05	−0.03	0.25[a]	1.00[a]												
	(0.463)	(0.533)	(0.725)	(0.001)	(0.000)												
Age	0.13[c]	−0.04	−0.10	0.10	−0.07	1.00[a]											
	(0.090)	(0.622)	(0.185)	(0.195)	(0.380)	(0.000)											
Sex	0.07	−0.05	0.07	−0.22[a]	−0.13[c]	0.12	1.00[a]										
	(0.351)	(0.525)	(0.325)	(0.003)	(0.087)	(0.115)	(0.000)										
PreviousExp	−0.51[a]	−0.07	0.06	−0.10	−0.07	−0.16[b]	−0.05	1.00[a]									
	(0.000)	(0.378)	(0.390)	(0.167)	(0.373)	(0.033)	(0.525)	(0.000)									
FinanceKnow	−0.02	0.25[a]	−0.06	0.38[a]	0.35[a]	0.11	−0.33[a]	−0.03	1.00[a]								
	(0.795)	(0.001)	(0.433)	(0.000)	(0.000)	(0.142)	(0.000)	(0.675)	(0.000)								
TradedStock	0.04	0.13[c]	−0.21[a]	0.15[b]	0.16[b]	0.25[a]	−0.27[a]	−0.06	0.53[a]	1.00[a]							
	(0.598)	(0.071)	(0.004)	(0.040)	(0.031)	(0.001)	(0.000)	(0.429)	(0.000)	(0.000)							
TradedOptions	0.03	0.02	−0.18[b]	−0.01	0.01	0.16[b]	−0.27[a]	−0.04	0.23[a]	0.38[a]	1.00[a]						
	(0.687)	(0.822)	(0.012)	(0.936)	(0.894)	(0.030)	(0.000)	(0.637)	(0.002)	(0.000)	(0.000)						
Result	0.05	0.48[a]	0.11	0.26[a]	−0.11	−0.08	−0.27[a]	−0.08	0.21[a]	0.19[a]	0.15[b]	1.00[a]					

	(0.488)	(0.000)	(0.122)	(0.000)	(0.121)	(0.277)	(0.000)	(0.255)	(0.004)	(0.009)	(0.048)	(0.000)					
StartingStock	0.00	0.10	0.06	0.03	−0.05	0.10	−0.07	0.03	0.04	0.01	0.00	0.06	1.00^{a}				
	−1.000	(0.176)	(0.394)	(0.672)	(0.511)	(0.170)	(0.340)	(0.699)	(0.606)	(0.853)	−1.000	(0.382)	(0.000)				
Education	0.21^{a}	0.08	−0.16^{b}	0.20^{a}	0.17^{b}	0.54^{a}	−0.05	−0.25^{a}	0.31^{a}	0.30^{a}	0.15^{b}	0.03	0.08	1.00^{a}			
	(0.005)	(0.298)	(0.029)	(0.005)	(0.021)	(0.000)	(0.473)	(0.001)	(0.000)	(0.000)	(0.037)	(0.726)	(0.298)	(0.000)			
StockTrans	0.01	−0.04	−0.08	0.16^{b}	0.12	0.08	−0.10	0.01	0.17^{b}	0.15^{b}	−0.03	−0.04	−0.03	0.07	1.00^{a}		
	(0.938)	(0.573)	(0.311)	(0.035)	(0.107)	(0.308)	(0.181)	(0.899)	(0.020)	(0.044)	(0.726)	(0.570)	(0.734)	(0.324)	(0.000)		
OptionTrans	0.00	−0.14^{c}	−0.11	−0.04	−0.02	0.10	−0.23^{a}	−0.10	0.26^{a}	0.23^{a}	0.30^{a}	0.08	−0.06	0.20^{a}	0.28^{a}	1.00^{a}	
	(0.952)	(0.051)	(0.155)	(0.577)	(0.833)	(0.182)	(0.002)	(0.171)	(0.000)	(0.001)	(0.000)	(0.257)	(0.446)	(0.006)	(0.000)	(0.000)	
Treatment	−0.14^{c}	−0.09	0.03	0.01	0.05	−0.12	−0.16^{b}	0.14^{c}	−0.07	−0.09	−0.09	0.05	0.00	−0.41^{a}	−0.08	−0.06	1.00^{a}
	(0.062)	(0.246)	(0.679)	(0.907)	(0.530)	(0.118)	(0.025)	(0.067)	(0.343)	(0.230)	(0.237)	(0.494)	−1.000	(0.000)	(0.289)	(0.412)	(0.000)

^{a}Significant at the 1%-level, ^{b}significant at the 5%-level, ^{c}significant at the 10%-level

This table lists Pearson correlations between various subject characteristics and performance indicators. The numbers in parentheses are *p*-values

The categories are:

Round................The experimental round ({1, 2, 3} in the DO8 treatment, {1, 2} in the DO5/10/15 treatment).
Rational.............Answer on question 4 of the questionnaire (see Table 12)
OMktHelped ......Answer on question 6 of the questionnaire (see Table 12)
Understanding....Answer on question 9 of the questionnaire (see Table 12)
Instructions.........Answer on question 10 of the questionnaire (see Table 12).
Age....................Answer on question 13 of the questionnaire (see Table 12)
Sex.....................Answer on question 14 of the questionnaire (see Table 12)
PreviousExp ....Answer on question 16 of the questionnaire, 0 if no prior experimental experience, 1 if prior experience (see Table 12)
FinanceKnow .....Answer on question 17 of the questionnaire (see Table 12)
TradedStock ....Answer on question 18 of the questionnaire (see Table 12)
TradedOptions ...Answer on question 19 of the questionnaire (see Table 12)
Result ....Terminal wealth within the experimental round, relative to mean subject terminal wealth, excluding the show-up fee and before application of the non-negativity clause
StartingStock......Size of initial stock endowment (see Table 3)
Education ....Answer on question 15 of the questionnaire (see Table 12)
StockTrans ....Number of stock transactions by this subject within the experimental round
OptionTrans ....Number of option transactions by this subject within the experimental round.
Treatment...........0 in the DO8 treatment, 1 in the DO5/10/15 treatment

Bibliography

What we become depends on what we read after all of the professors have finished with us. The greatest university of all is a collection of books.

Thomas Carlyle, 1795–1881

Ackert, L.F. Church, B.K.: The effects of subject pool and design experience on rationality in experimental asset markets. J. Psychol. Financ. Market **2**(1), 6–28 (2001)

Ackert, L.F. Church, B.K. Jayaraman, N.: An experimental study of circuit breakers: the effects of mandated market closures and temporary halts on market behavior. J. Financ. Market. **4**, 185–208 (2001)

Ackert LF, Charupat N, Deaves R, Kluger BD. The origins of bubbles in laboratory asset markets. Working Paper, Federal Reserve Bank of Atlanta 2006-6 (2006c)

Ackert, L.F. Charupat, N. Church, B.K. Deaves, R.: Margin, short selling, and lotteries in experimental asset markets. South. Econ. J. **73**(2), 419–436 (2006b)

Ackert, L.F. Charupat, N. Church, B.K. Deaves, R.: An experimental examination of the house money effect in a multi-period setting. Exp. Econ. **9**, 5–16 (2006c)

Alevy, J.E. Haigh, M.S. List, J.A.: Information cascades: evidence from a field experiment with financial market professionals. J. Financ. **62**(1), 151–180 (2007)

Allen, F. Gorton, G.: Churning bubbles. Rev. Econ. Stud. **60**, 813–836 (1993)

Allen, F. Morris, S. Postlewaite, A.: Finite bubbles with short sale constraints and asymmetric information. J. Econ. Theory **61**, 206–229 (1993)

Amihud, Y.: Bidding and auctioning for procurement and allocation. New York University Press, New York (1976)

Amihud, Y. Mendelson, H.: Asset pricing and the bid-ask spread. J. Financ. Econ. **17**, 223–250 (1986)

Amihud, Y. Mendelson, H.: Liquidity, asset prices, and financial policy. Financ. Analysts. J. **47**, 56–66 (1991)

Ang, J.S., Diavatopoulos, D., Schwarz, T.: The creation and control of speculative bubbles in a laboratory setting. Florida State University and Grand Valley State University, Working Paper (1992)

Arthur, B.W., Holland, J.H., LeBaron, B., Palmer, R., Tayler, P.: Asset pricing under endogenous expectations in an artificial stock market. Working Paper, doi: 10.2139, ssrn.2252 (1996)

Bachelier, L.: Théorie de la Speculation. Gauthier-Villars, Paris (1900)

Barlevy, G.: Economic theory and asset bubbles. Econ. Perspec **3rd Quarter**, 44–59 (2007)

Berg, J., Nelson, F., Rietz, T.: Accuracy and forecast standard error of prediction markets. Working Draft (2003)

Bernoulli, D.: Specimen Theoriae novae de mensura sortis. Commentarii Academiae Scientiarum Imperialis Petropolitanae **5**, 175–192 (1738). English translation in Econometrica 22, 23-36 (1954)

Berry, T.D. Howe, K.M.: Public information arrival. J. Financ. **49**(4), 1331–1346 (1994)

Biais, B. Hillion, P.: Insider and liquidity trading in stock and options markets. Rev. Financ. Stud. **7**(4), 743–780 (1994)

Black, F.: Noise. J. Financ. **16**, 529–543 (1986)

Black, F. Scholes, M.: The pricing of options and corporate liabilities. J. Polit. Econ. **81**, 637–654 (1973)

Bodie, Z. Kane, A. Marcus, A.J.: Investments, 6 International Edition. McGraw-Hill, New York (2005)

Caginalp, G. Porter, D. Smith, V.L.: Initial cash, asset ratio and asset prices: an experimental study. Proc. Natl. Acad. Sci. USA **95**, 756–761 (1998)

Caginalp, G. Porter, D. Smith, V.L.: Momentum and overreaction in experimental asset markets. Int. J. Indust. Organ **18**, 187–204 (2000a)

Caginalp, G. Porter, D. Smith, V.L.: Overreactions, momentum, liquidity, and price bubbles in laboratory and field asset markets. J. Psychol. Financ. Market **1**(1), 24–48 (2000b)

Caginalp, G. Porter, D. Smith, V.L.: Financial bubbles: excess cash, momentum, and incomplete information. J. Psychol. Financ. Market. **2**(2), 80–99 (2001)

Caginalp, G. Ilieva, V. Porter, D. Smith, V.L.: Do speculative stocks lower prices and increase volatility of value stocks? J. Psychol. Financ. Market **3**(2), 118–132 (2002)

Camerer, C.F.: Do biases in probability judgment matter in markets? Experimental evidence. Am Econ Rev **77**(5), 981–997 (1987)

Camerer, C.F.: Bubbles and fads in asset prices. J. Econ. Surv. **3**(1), 3–41 (1989)

Camerer, C.F. Hogarth, R.M.: The Effects of financial incentives in experiments: a review and capital-labor-production framework. J. Risk. Uncert. **19**(1–3), 7–42 (1999)

Camerer, C.F. Weigelt, K.: Information mirages in experimental asset markets. J. Bus. **64**(4), 463–493 (1991)

Cao, H.H.: The effect of derivative assets an information acquisition and price behavior in a rational expectations equilibrium. Rev. Financ. Stud. **12**, 131–163 (1999)

Carhart, M.M.: On persistence in mutual fund performance. J. Financ. **52**(1), 57–82 (1997)

Carley, K. Prietula, M. (eds.): Computational organization theory. Lawrence Erlbaum Associates, Hillsdale, New Jersey (1994)

Cason, T.N. Friedman, D.: Price formation in double auction markets. J. Econ. Dyn. Control **20**, 1307–1337 (1996)

Cass, D. Shell, K.: Do sunspots matter? J. Polit. Econ. **91**(2), 193–227 (1983)

Chakravarty, S. Gulen, H. Mayhew, S.: Informed trading in stock and option markets. J. Financ. **59** (3), 1235–1257 (2004)

Chamberlin, E.H.: An experimental imperfect market. J. Polit. Econ. **56**(2), 95–108 (1948)

Chen, K. Fine, L.R. Huberman, B.A.: Predicting the future. Inf. Syst. Front **5**(1), 47–61 (2003)

Chern, K.-Y. Tandon, K. Yu, S. Webb, G.: The information content of stock split announcements: do options matter? J. Bank. Financ. **32**(6), 930–946 (2008)

Chordia, T. Roll, R. Subrahmanyam, A.: Commonality and liquidity. J. Financ. Econ. **56**, 3–28 (2000)

Conrad, J. Kaul, G.: Time-variation in expected returns. J. Bus. **61**, 409–425 (1988)

Cootner, P. (ed.): The Random Character of Stock Market Prices. MIT, Cambridge (1964)

Corgnet, B. Kujal, P. Porter, D.: Uninformative Announcements and Asset Trading Behavior. Universidad Carlos III De Madrid, Working Paper, Madrid (2007)

Corgnet, B., Kujal, P., Porter, D.: Asset trading behavior under reliable and unreliable public statements. Working Paper (2008)

Cox, C.C.: Futures trading and market information. J. Polit. Econ. **84**(6), 1215–1237 (1976)

Cross, F.: The behavior of stock market prices on fridays and mondays. Financ. Analysts J. **29**(6), 67–69 (1973)

Crowley, S. Sade, O.: Does the option to cancel an order in a double auction market matter? Econ. Lett. **83**, 89–97 (2004)

Davies, T.: Irrational Gloominess in the Laboratory. University of Arizona, Arizona (2006). Working Paper

Davis, D.D. Holt, C.A.: Experimental Economics. Princeton University Press, Princeton, New Jersey (1993)

Day, R. Chen, P. (eds.): Nonlinear Dynamics and Evolutionary Economics. Oxford University Press, Oxford (1993)

De Bondt, W.F.M. Thaler, R.: Does the stock market overreact? J. Financ. **15**, 793–805 (1985)

De Jong, C. Koedijk, K.G. Schnitzlein, C.R.: Stock market quality in the presence of a traded option. J. Bus. **79**(4), 2243–2274 (2006)

De Long, J.B. Shleifer, A. Summers, L.H. Waldmann, R.J.: Positive feedback investment strategies and destabilizing rational speculation. J. Financ. **45**(2), 379–395 (1990)

Diba, B.T. Grossman, H.I.: Explosive rational bubbles in stock prices? Am. Econ. Rev. **78**(3), 520–530 (1988)

Donaldson, R.G. Kamstra, M.: A new dividend forecasting procedure that rejects bubbles in asset prices: the case of 1929's stock crash. Rev. Financ. Stud. **9**(2), 333–383 (1996)

Dufwenberg, N. Lindqvist, T. Moore, E.: Bubbles and experience: an experiment. Am. Econ. Rev. **95**(5), 1731–1737 (2005)

Dyer, D. Kagel, J.H. Levin, D.: A comparison of naïve and experienced bidders in common value offer auctions: a laboratory analysis. Econ. J. **99**, 108–115 (1989)

Easley, D., Ledyard, J.O.: Theories of price formation and exchange in double oral auctions. In: Friedman, D., Geankoplos, J., Lane, D., Rust, J. (eds.) The Double Auction Market: Institutions, Theories and Evidence. Santa Fe Institute Studies in the Sciences of Complexity, Proceedings, vol. 15. Addison-Wesley, Reading, MA (1992)

Easley, D. O'Hara, M. Srinivas, P.S.: Option volume and stock prices: evidence on where informed traders trade. J. Financ. **53**, 431–465 (1998)

Eckel, C.C. Grossman, P.J.: Volunteers and pseudo-volunteers: the effect of recruitment method in dictator experiments. Exp. Econ. **3**, 107–120 (2000)

Faff, R.W., Hallahan, T., McKenzie, M.D.: An empirical investigation of personal financial risk tolerance. Working Paper (2008)

Fama, E.: Efficient capital markets: A review of theory and empirical work. J. Financ. **25**, 384–417 (1970)

Fama, E.: Efficient capital markets: II. J. Financ. **46**, 1575–1617 (1991)

Fama, E.: Market efficiency, long-term returns, and behavioral finance. J. Financ. Econ. **49**, 283–306 (1998)

Fama, E.F. French, K.R.: Permanent and temporary components of stock prices. J. Polit. Econ. **96**, 246–273 (1988)

Fama, E.F. French, K.R.: Common risk factors in the returns on stocks and bonds. J. Financ. Econ. **33**(1), 3–56 (1993)

Figlewski, S. Webb, G.: Options, short sales, and market completeness. J. Financ. **48**, 761–777 (1993)

Fischbacher, U.: z-Tree: Zurich toolbox for ready-made economic experiments. Exp. Econ. **10**(2), 171–178 (2007)

Fisher, E.O'.N. Kelly, F.S.: Experimental foreign exchange markets. Pac. Econ. Rev. **5**(3), 365–387 (2000)

Forsythe, R. Palfrey, T.R. Plott, C.R.: Asset valuation in an experimental market. Econometrica **50** (3), 537–567 (1982)

Forsythe, R. Palfrey, T.R. Plott, C.R.: Futures markets and informational efficiency: A laboratory examination. J. Financ. **39**(4), 955–981 (1984)

Forsythe, R. Nelson, F. Neumann, G.R. Wright, J.: Anatomy of an experimental political stock market. Am. Econ. Rev. **82**(5), 1142–1161 (1992)

French, K.: Stock returns and the weekend effect. J. Financ. Econ. **8**, 55–69 (1980)

Friedman, B.: Stock prices and social dynamics: comments and discussion. Brooking Pap. Econ. Act. **2**, 504–508 (1984a)

Friedman, D.: On the efficiency of experimental double auction markets. Am. Econ. Rev. **74**(1), 60–72 (1984b)

Friedman, D. Harrison, G.W. Salmon, J.W.: The informational role of futures markets: some experimental evidence. In: Streit, M.E. (ed.) Futures Markets – Modelling. Managing and Monitoring Futures Trading. Basil Blackwell, Oxford (1983)

Friedman, D. Harrison, G.W. Salmon, J.W.: The informational efficiency of experimental asset markets. J. Polit. Econ. **92**(3), 349–408 (1984)

Friedman, D., Geankoplos, J., Lane, D., Rust, J. (eds.): The Double Auction Market: Institutions, Theories and Evidence. Santa Fe Institute Studies in the Sciences of Complexity, Proceedings, vol. 15. Addison-Wesley, Reading, MA (1992)

Gan, J.: The real effects of asset market bubbles: loan- and firm-level evidence of a lending channel. Rev. Financ. Stud. **20**(5), 1941–1973 (2007)

Gibbons, M. Hess, P.: Day of the week effects and asset returns. J. Bus. **54**, 579–596 (1981)

Gode, D.K. Sunder, S.: Allocative efficiency of markets with zero-intelligence traders: market as a partial substitute for individual rationality. J. Polit. Econ. **101**(1), 119–137 (1993)

Gode, D.K. Sunder, S.: Human and artificially intelligent traders in a double auction market: experimental evidence. In: Carley, K. Prietula, M. (eds.) Computational Organization Theory. Lawrence Erlbaum Associates, Hillsdale, New Jersey (1994)

Goethe, V.J.W.: Faust II. Reclam, Ditzingen (1986)

Greenspan, A.: The challenge of central banking in a democratic society. http://www federalreserve. gov/boarddocs/speeches/1996/19961205.htm, December 5, 1996, Accessed. Aug 11, 2008

Grossman, S.J.: On the efficiency of competitive stock markets where trades have diverse information. J. Financ. **31**(2), 573–585 (1976)

Grossman, S.J.: An analysis of the implications for stock and futures price volatility of program trading and dynamic hedging strategies. J. Bus. **61**(3), 275–298 (1988)

Grossman, S.J. Stiglitz, J.E.: On the impossibility of informationally efficient markets. Am. Econ. Rev. **70**, 393–408 (1980)

Guenster. N., Kole, E., Jacobsen, B.: Riding bubbles. Working Paper (2007)

Guth, W. Krahnen, J.P. Rieck, C.: Financial markets with asymmetric information: a pilot study focusing on insider advantages. J. Econ. Psychol. **18**, 235–257 (1997)

Haase, E.E.: Harmonisierung der Rechnungslegung - eine experimentelle Untersuchung der Auswirkung der IAS-Verordnung auf den Kapitalmarkt in Österreich, unpublished dissertation, Karl-Franzens-University Graz (2006)

Harrison, J.M. Kreps, D.M.: Speculative investor behavior in a stock market with heterogeneous expectations. Q. J. Econ. **92**(2), 323–336 (1978)

Haruvy, E. Noussair, C.N.: The effect of short selling on bubbles and crashes in experimental spot asset markets. J. Financ. **61**(3), 1119–1157 (2006)

Haruvy, E. Lahav, Y. Noussair, C.N.: Traders' expectations in asset markets: experimental evidence. Am. Econ. Rev. **97**(5), 1901–1920 (2007)

Hasbrouck, J.: One security, many markets: Determining the contributions to price discovery. J. Financ. **50**(4), 1175–1199 (1995)

Hawawini, G.A.: European equity markets: a Review of the evidence on price behaviour and efficiency. In: Hawawini, G.A. Michel, P.A. (eds.) European equity markets-risk return and efficiency. Garland Publishing, New York (1984)

Hawawini, G.A. Michel, P.A. (eds.): European equity markets-risk return and efficiency. Garland Publishing, New York (1984)

Hertwig, R. Ortman, A.: Experimental Practices in Economics: A Challenge for Psychologists? Max Planck Institute for Human Development, Berlin (2001). Working Paper

Hirota, S. Sunder, S.: Price bubbles sans dividend anchors: evidence from laboratory stock markets. J. Econ. Dyn. Control **31**, 1875–1909 (2007)
Holt, C.A.: Industrial organization: a survey of laboratory research. In: Kagel, J.H. Roth, A.E. (eds.) Handbook of experimental economics. Princeton University Press, Princeton, New Jersey (1995)
Huber, J. Kirchler, M. Stockl, T.: Bubble or no bubble – the impact of model design on the formation of price bubbles in experimental asset markets. University of Innsbruck, Austria (2008). Working Paper
Hussam, R.N. Porter, D. Smith, V.L.: Thar she blows: can bubbles be rekindled with experienced subjects? Am. Econ. Rev. **98**(3), 924–937 (2008)
IMF: World economic outlook: housing and the business cycle, IMF World Economic Outlook, http://www.imf.org/external/pubs/ft/weo/2008/01/pdf/text.pdf, April 2008, Accessed 17 Aug 2008
Independent: Niall Ferguson: A stock market bubble is nothing new. http://www.independent.co.uk/opinion/commentators/niall-ferguson-a-stock-market-bubble-is-nothing-new-689055.html, 26 March 2001, Accessed 17 Sept 2008
International Herald Tribune: Real estate bubbles: how worried should we be? http://www.iht.com/articles/2007/01/12/business/wbreal.php, 12 Jan 2007, Accessed 17 Sept 2008
Isaac, R.M. (ed.): Research in experimental economics, vol. 5. JAI, Greenwich, CT (1992)
Jackson, M.O. Swinkels, J.M.: Existence of equilibrium in single and double private value auctions. Econometrica **73**(1), 93–139 (2005)
Jaffe, J.F.: Special information and insider trading. J. Bus. **47**, 410–428 (1974)
James, D. Isaac, R.M.: Asset markets: how they are affected by tournament incentives for individuals Am. Econ. Rev. **90**(4), 995–1004 (2000)
Jayaraman, N. Frye, M.B. Sabherwal, S.: Informed trading around merger announcements: an empirical test using transaction volume and open interest in options markets. Financ. Rev. **37**, 45–74 (2001)
Jegadeesh, N. Titman, S.: Returns to buying winners and selling losers: implications for stock market efficiency. J. Financ. **48**, 65–91 (1993)
Jegadeesh, N. Titman, S.: Profitability of momentum strategies: an evaluation of alternative explanations. J. Financ. **56**, 699–720 (2001)
Jennings, R. Starks, L.: Earnings announcements, stock price adjustment, and the existence of option markets. J. Financ. **41**(1), 107–125 (1986)
Kagel, J.H. Roth, A.E. (eds.): Handbook of experimental economics. Princeton University Press, Princeton, New Jersey (1995)
Keim, D.B.: Size related anomalies and stock return seasonality: further empirical evidence. J. Financ. Econ. **12**, 13–32 (1983a)
Keim, D.B.: Further evidence on size effects and yield effects: the implications of stock return seasonality. University of Chicago, Chicago, IL (1983b). Unpublished Manuscript
Keim, D.B. Stambaugh, R.F.: A further investigation of the weekend effect in stock returns. J. Financ. **39**, 819–835 (1984)
Kendall, M.: The analysis of economic time series. Part I: Prices. J. R. Stat. Soc **116**, 11–34 (1953)
Keynes, J.M.: The General Theory of Employment, Interest and Money. Harcourt Brace, New York (1936)
King, R.R.: Private information acquisition in experimental markets prone to bubble and crash. J. Financ. Res. **14**(3), 197–206 (1991)
King, R.R. Smith, V.L. Williams, A.W. Van Boening, M.: The robustness of bubbles and crashes in experimental stock markets. In: Day, R. Chen, P. (eds.) Nonlinear Dynamics and Evolutionary Economics. Oxford University Press, Oxford (1993)
Kluger, B.D. Wyatt, S.B.: Options and efficiency: some experimental evidence. Rev. Quant. Financ. Account. **5**, 179–201 (1995)
Koessler, F., Noussair, C., Ziegelmeyer, A.: Individual behavior and beliefs in experimental parimutuel betting markets. Unpublished Working Paper (2005)

Kraus, A. Smith, M.: Endogenous sunspots, pseudo-bubbles, and beliefs about beliefs. J. Financ. Market. **1**, 151–174 (1998)

Lee, J. Yi, C.H.: Trade size and information-motivated trading in the options and stock markets. J. Financ. Quant. Anal. **36**(4), 485–501 (2001)

Lei, V. Noussair, C.N. Plott, R.: Nonspeculative bubbles in experimental asset markets: lack of common knowledge of rationality vs. actual irrationality. Econometrica **69**(4), 831–859 (2001)

Leitner, J. Schmidt, R.: A systematic comparison of professional exchange rate forecasts with the judgemental forecasts of novices. Cent. Eur. J. Oper. Res. **14**(1), 87–102 (2006)

Lintner, J.: The valuation of risk assets and the selection of risky investments in stock portfolios and capital budgets. Rev. Econ. Stat. **47**, 13–37 (1965)

Liu, Y.-J.: Auction mechanisms and information structure: an experimental study of information aggregation in securities markets. In: Isaac, pp. 165–212 (1992) (as quoted in Sunder (1995), p. 486)

Lorie, J.H. Brealey, R.: Modern Developments in Investment Management, 2nd edn. Dryden, Hinsdale, IL (1978)

Lorie, J.H. Niederhoffer, V.: Predictive and statistical properties of insider trading. J. Law Econ. **11**(1), 35–53 (1968)

Los Angeles Times: Disappearing now: $6 trillion in housing wealth http://latimesblogs.latimes.com/laland/2008/04/disappearing-no.html, 30 Apr 2008, Accessed 02 Aug 2008

Los Angeles Times: If 'Bubble' Bursts, Legacy of Greenspan May Deflate. http://articles.latimes.com/2005/aug/26/business/fi-fed26, 26 Aug 2005. Accessed: 17 Sept 2008

Lowry, S.T.: Ancient and medieval economics. In: Samuels, W.J. Biddle, J.E. Davis, J.B. (eds.) The History of Economic Thought. Blackwell, Malden, MA (2007)

Luckner, S. Weinhardt, C.: How to pay traders in information markets: results from a field experiment. J. Prediction Market **1**(2), 147–156 (2007)

Malkiel, B.G. Mullainathan, S. Stangle, B.: Market efficiency versus behavioral finance. J. Appl. Corp. Financ. **17**(3), 124–134 (2005)

Manaster, S. Rendleman, R.J.: Option prices as predictors of equilibrium stock prices. J. Financ. **37** (4), 1043–1057 (1982)

Manski, C.: Interpreting the Predictions of Prediction Markets. NBER Working Paper #10359, March 2004

Marsh, T.A. Merton, R.C.: Dividend variablility and variance bounds tests for the rationality of stock market prices. Am. Econ. Rev. **76**, 483–498 (1986)

Milgrom, P.R. Weber, R.J.: A theory of auctions and competitive bidding. Econometrica **50**(5), 1089–1122 (1982)

Miller, R.M.: Can markets learn to avoid bubbles? J. Psychol. Financ. Market **3**(1), 44–52 (2002)

Mossin, J.: Equilibrium in a capital asset market. Econometrica **34**, 768–783 (1966)

Moulin, H.: Game theory for the social sciences, 2nd edn. New York University Press, New York (1986)

Muth, J.F.: Rational expectations and the theory of price movements. Econometrica **29**(3), 315–335 (1961)

Nash, J.F.: The bargaining problem. Econometrica **18**, 155–162 (1950)

New York Times: Two bubbles, two paths. http://www.nytimes.com/2008/06/15/business/15view.html, 15 Jun 2008, Accessed 17 Sept 2008

Noussair, C.N. Powell, O.: Peaks and Valleys: Experimental Asset Markets with Non-Monotonic Fundamentals. Tilburg University, The Netherland (2008). Working Paper

Noussair, C.N. Tucker, S.: Futures markets and bubble formation in experimental asset markets. Pac. Econ. Rev. **11**(2), 167–184 (2006)

Noussair, C.N. Robin, S. Ruffieux, B.: Price bubbles in laboratory asset markets with constant fundamental values. Exp. Econ. **4**, 87–105 (2001)

NYSE, NYSE Rules: http://rules.nyse.com/NYSE/NYSE_Rules/ (2008). Accessed 9 Jan 2008

O'Hara, M.: Market Microstructure Theory. Blackwell, Cambridge, MA (1995)

Oechssler, J., Schmidt, C., Schnedler, W.: Asset bubbles without dividends – an experiment. University of Mannheim. Working Paper 07-01 (2007)

Oldfield, G.S. Rogalski, R.J.: A theory of common stock returns over trading and non-trading periods. J. Financ. **35**(3), 729–751 (1980)

Oliver, P.: Financial binary betting, styles, valuations and deductions from data. J. Prediction Market. **1**(2), 127–146 (2007)

Ortner, G.: Qualitätskriterien der Informationsverarbeitung in Borsen am Beispiel Political Stock Markets. University of Vienna, Austria (1996). Unpublished Dissertation

Palan, S.: The efficient market hypothesis and its validity in today's markets. Grin, Munich (2004)

Pan, J., Poteshman, A.M.: The information in option volume for stock prices. MIT Sloan Working Paper No. 4276-03 (2003)

Pástor, L. Veronesi, P.: Was there a Nasdaq bubble in the late 1990s? J. Financ. Econ. **81**, 61–100 (2006)

Peterson, S.P.: Forecasting dynamics and convergence to market fundamentals. J. Econ. Behav. Organ. **22**, 269–284 (1993)

Plott, C.R.: Industrial organization theory and experimental economics. J. Econ. Lit. **20**, 1485–1527 (1982)

Plott, C.R.: Rational expectations and the aggregation of diverse information in laboratory security markets. Econometrica **56**(5), 1085–1118 (1988)

Plott, C.R.: Will economics become an experimental science? South. Econ. J. **57**(4), 901–919 (1991)

Plott, C.R.: Markets as information gathering tools. South. Econ. J. **67**(1), 1–15 (2000)

Plott, C.R. Sunder, S.: Efficiency of experimental security markets with insider information: an application of rational-expectations models. J. Polit. Econ. **90**(41), 663–698 (1982)

Porter, D.P. Smith, V.L.: Stock market bubbles in the laboratory. Appl. Math. Financ. **1**, 111–127 (1994)

Porter, D.P. Smith, V.L.: Futures contracting and dividend uncertainty in experimental asset markets. J. Bus. **68**(4), 509–541 (1995)

Poterba, J. Summers, L.: Mean reversion in stock prices: evidence and implications. J. Financ. Econ. **22**, 27–59 (1988)

Pouget, S.: Financial market design and bounded rationality: an experiment. J. Financ. Market. **10**, 287–317 (2007)

Pratt, S.P.: Valuing a Business: The Analysis of Closely Held Companies, 2nd edn. Homewood, IL (1989)

Pratt, S.P., DeVere, C.W.: Relationship between insider trading and rates of return for NYSE common stocks, 1960-66. Unpublished paper presented to the seminar on the analysis of security prices, May 1968, University of Chicago in Lorie and Brealey, pp. 259–270 (1978)

Ritter, J.R.: The buying and selling behaviour of individual investors at the turn of the year. J. Financ. **43**, 701–717 (1988)

Ritter, J.R. Chopra, N.: Portfolio rebalancing and the turn-of-the-year effect. J. Financ. **44**, 149–166 (1989)

Rogalski, R.J.: Discussion of Keim and Stambaugh (1984). J. Financ. **39**, 835–837 (1984a)

Rogalski, R.J.: New findings regarding day-of-the-week returns over trading and non-trading periods: a note. J. Financ. **39**, 1603–1614 (1984b)

Roll, R.: A critique of the asset pricing theory's tests. J. Financ. Econ. **4**(2), 129–176 (1977)

Roll, R.: The hubris hypothesis of corporate takeovers. J. Bus. **59**(2), 197–216 (1986)

Ross, S.A. Westerfield, R.W. Jaffe, J.F.: Corporate Finance, 7 International edn. McGraw-Hill, New York (2005)

Roth, A.E.: Introduction to experimental economics. In: Kagel, J.H. Roth, A.E. (eds.) Handbook of Experimental Economics. Princeton University Press, Princeton, New Jersey (1995)

Rouwenhorst, K.G.: International momentum strategies. J. Financ. **53**, 267–284 (1998)

Royal Swedish Academy of Sciences: The sveriges riksbank prize in economic sciences in memory of Alfred Nobel 2002. Press Release, http://nobelprize.org/nobel_prizes/economics/laureates/2002/press.html, October 09, 2002. Accessed May 10, 2008

Samuels, W.J., Biddle, J.E., Davis, J.B. (eds.). The History of Economic Thought. Blackwell, Malden (2007)

Schlag, C. Stoll, H.: Price impacts of options volume. J. Financ. Market **8**, 69–87 (2005)

Schwartz, R.A.: Discussion of Fama (1970). J. Financ. **25**, 421–423 (1970)

Seyhun, N.: Insiders' profits, costs of trading and market efficiency. J. Financ. Econ. **16**, 189–212 (1986)

Sharpe, W.F.: Capital asset prices: a theory of market equilibrium under conditions of risk. J. Financ. **19**(3), 416–422 (1964)

Shiller, R.J.: Do stock prices move too much to be justified by subsequent changes in dividends? Am. Econ. Rev. **71**, 421–436 (1981)

Shiller, R.J.: From efficient markets theory to behavioral finance. J. Econ. Perspect. **17**(1), 83–104 (2003)

Shleifer, A.: Do demand curves for stocks slope down? J. Financ. **41**, 579–590 (1986)

Shleifer, A. Summers, L.H.: The noise trader approach to finance. J. Econ. Perspect **4**, 19–33 (1990)

Siegel, J.J.: The future for investors: why the tried and the true triumph over the bold and the new. Crown Publishing, USA (2005)

Simon, H.A.: A behavioral model of rational choice. Q.J. Econ. **69**(1), 99–118 (1955)

Skinner, D.J.: Option markets and the information content of accounting earnings releases. J. Account. Econ. **13**, 191–211 (1990)

Smith, A.: An Inquiry Into the Nature and Causes of the Wealth of Nations, 3rd edn, p. 1843. Charles Knight, London (1843)

Smith, V.L.: An experimental study of competitive market behavior. J. Polit. Econ. **70**(2), 111–137 (1962)

Smith, V.L.: Effect of market organization on competitive equilibrium. Q. J. Econ. **78**(2), 181–201 (1964)

Smith, V.L.: Experimental auction markets and the walrasian hypothesis. J. Polit. Econ. **73**(4), 387–393 (1965)

Smith, V.L.: Experimental economics: induced value theory. Am. Econ. Rev. Pap. Proc. **66**(2), 274–279 (1976a)

Smith, V.L.: Bidding and auctioning institutions: experimental results. In: Amihud, Y. (ed.) Bidding and Auctioning for Procurement and Allocation. New York University Press, New York (1976b)

Smith, V.L.: Microeconomic systems as an experimental science. Am. Econ. Rev. **72**(5), 923–955 (1982)

Smith, V.L.: Experimental economics: reply. Am. Econ. Rev. **75**(1), 265–272 (1985)

Smith, V.L.: Economics in the laboratory. J. Econ. Perspect. **8**(1), 113–131 (1994)

Smith, V.L.: Notes on some literature in experimental economics." Social Science Working Paper, pp. 1–27, February (1973)

Smith, V.L. Walker, J.M.: Rewards, experience and decision costs in first price auctions. Econ. Inquiry **31**, 237–245 (1993a)

Smith, V.L. Walker, J.M.: Monetary rewards and decision cost in experimental economics. Econ. Inquiry **31**, 245–261 (1993b)

Smith, V.L. Williams, A.W.: On nonbinding price controls in a competitive market. Am. Econ. Rev. **71**(3), 467–474 (1981)

Smith, V.L. Williams, A.W. Bratton, W.K. Vannoni, M.G.: Competitive market institutions: double auctions vs. sealed bid-offer auctions. Am. Econ. Rev. **72**(1), 58–77 (1982)

Smith, V.L. Suchanek, G.L. Williams, A.W.: Bubbles, crashes and endogenous expectations in experimental spot asset markets. Econometrica **56**(5), 1119–1151 (1988)

Smith, V.L. van Boening, M. Wellford, C.P.: Dividend timing and behavior in laboratory asset markets. Econ. Theory **16**, 567–583 (2000)

Stanley, T.D.: Silly bubbles and the insensitivity of rationality testing: an experimental illustration. J. Econ. Psychol. **15**, 601–620 (1994)

Stanley, T.D.: Bubbles, inertia, and experience in experimental asset markets. J. Socio-Econ. **26** (6), 611–625 (1997)
Streit, M.E. (ed.): Futures Markets – Modelling. Managing and Monitoring Futures Trading. Basil Blackwell, Oxford (1983)
Sunder, S.: Experimental asset markets: a survey. In: Kagel/Roth, pp. 445–500 (1995)
Svenson, O.: Are we all less risky and more skillful than our fellow drivers are? 148 **47**, 143–148 (1981). quoted after Camerer (1989)
Tetlock, P.C.: How efficient are information markets? evidence from an online exchange. Harvard University, USA (2004). Working Paper
Theil, H.: Principles of Econometrics. Wiley, New York (1971)
Tremel, F.: Wirtschafts- und Sozialgeschichte O¨sterreichs. Franz Deuticke Verlag, Wien (1969)
Tversky, A. Kahneman, D.: Judgment under uncertainty: heuristics and biases. Science **185**, 1124–1131 (1974)
Tziralis, G. Tatsiopoulos, I.: Prediction markets: an extended literature review. J. Prediction Market **1**(1), 75–91 (2007)
Van Boening, M.V. Williams, A.W. LaMaster, S.: Price bubbles and crashes in experimental call markets. Econ. Lett. **41**, 179–185 (1993)
Von Békésy, G.: Experiments in hearing. McGraw-Hill Education, New York (1960)
Walter, R.: Geschichte der Weltwirtschaft. Bohlau Verlag, Koln (2006)
Westerhoff, F.: Bubbles and crashes: optimism, trend extrapolation and panic. Int. J. Theor. Appl. Financ. **6**(8), 829–837 (2003)
Williams, A.W.: Computerized double auction markets: some initial experimental results. J. Bus. **53**(3), 235–258 (1980)
Williams, A.W.: The formation of price forecasts in experimental markets. J. Money Credit Bank **19**(1), 18 (1987)
Williams, A.W.: Price bubbles in large financial asset markets. In: Plott, C. Smith, V.L. (eds.) Handbook of Experimental Economic Results, vol. 1, 1st edn, pp. 242–246. North-Holland, Amsterdam (2008)
Williams, A.W. Walker, J.M.: Computerized laboratory exercises for microeconomics education: three applications motivated by experimental economics. J. Econ. Educ. **24**(4), 291–315 (1993)
Wilson, R.: A bidding model of perfect competition. Rev. Econ. Stud. **44**(3), 511–518 (1977)
Wolfers, J. Zitzewitz, E.: Prediction markets. J. Econ. Perspect **18**(2), 107–126 (2004)
Wolfers, J., Zitzewitz, E.: Interpreting prediction market prices as probabilities. NBER Working Paper (2006)
Ziemba, W.T.: World wide security market regularities. Eur. J. Oper Res. **74**(2), 198–229 (1994)

Index

Breinigsville, PA USA
01 February 2010
231612BV00002BA/8/P

9 783642 021466